FUNDAMENTALS OF
WIND ENERGY

By
NICHOLAS P. CHEREMISINOFF

ANN ARBOR SCIENCE
PUBLISHERS INC
P.O. BOX 1425 • ANN ARBOR, MICH. 48106

Second Printing, 1978

Copyright © 1978 by Ann Arbor Science Publishers, Inc.
230 Collingwood, P. O. Box 1425, Ann Arbor, Michigan 48106

Library of Congress Catalog Card No. 78-51051
ISBN 0-250-40255-6

Manufactured in the United States of America
All Rights Reserved

PREFACE

Today's energy sources, oil and natural gas, are running out. Can tomorrow's energy come from the wind? Some think so, while others believe that the atmosphere is too vast and that we cannot harness wind power on a major scale. Wind, blowing constantly in swift currents around the world, offers enormous energy potential. In spite of man's ability to create mighty instruments, he has not been able to harness totally this awesome power of nature. The wind remains beyond a desirable level of control in spite of all our technological advances.

Converting the forces of the winds into useful energy intrigues the experts as its potential dazzles with promise. Attention is being focused on the prospects with the hope that progress can be made in developing the means to control these savage forces and channel them into useful energy production. In what time frame can this be done—years? decades?—and how does the wind relate to all of the other energy possibilities being investigated?

The idea of using the wind as a source of power or energy to do man's bidding is not new. The earliest known wind machines, used by the Persians to grind grain, predate Christian civilization. Use of windmills is recorded in the 12th century in France and England and was extensive throughout Europe by the 13th century. Wind machines and the typical American windmill were employed in developing agriculture and local water supply in our own West.

And so, beset with rising prices, environmental disadvantages and depleting reserves of 20th-century energy sources, the fossil fuels, the technologist/engineer is looking at energy alternates.

This book is presented as an overview of the potentials of wind as an energy source. It is intended as a documentary offering a look at the past as well as at the present and future possibilities.

Nicholas P. Cheremisinoff

NICHOLAS P. CHEREMISINOFF received his B.S., M.S. and Ph.D. degrees in chemical engineering from Clarkson College of Technology where he was instructor during 1976-77. He is currently Research Scientist at Union Camp Corp. R & D Division in Princeton, New Jersey. Dr. Cheremisinoff has contributed to industrial press and technical publications and has authored sections in several engineering handbooks. He is a member of several professional and honorary societies including Tau Beta Pi and Sigma Xi, and he has received numerous honors and awards.

CONTENTS

1. Wind Energy—Overview 1
2. History of Wind Energy 15
3. Modern Applications of Wind Energy 29
4. Wind Machines and Generators 49
5. Performance and Design Characteristics 69
6. Wind Site Selection Factors 85
7. Energy Storage Systems and Environmental Considerations 99
8. Future Potential of Wind Energy 123
Appendix A. Glossary of Energy-Related Terms 129
Appendix B. Conversion Factors 139
References 159
Index 165

CHAPTER 1

WIND ENERGY—OVERVIEW

"Every great advance in science has issued from a new audacity of imagination." . . . . John Dewey

Winds are the motion of air about the earth caused by its rotation and by the uneven heating of the planet's surface by the sun. During the daytime, the air over the earth's crust acts partly as an absorber and partly as a reflector. That is, some of the sun's energy is absorbed by the land, but a larger portion is reflected back, heating the atmosphere. Over the lakes and oceans, a great deal of this energy is absorbed by water or is involved in evaporation; hence, this air remains relatively cool. The warmed air over the land expands, becomes lighter and rises, causing the heavier, cooler air over the bodies of water to move in and replace it. Thus, local shoreline breezes are merely a displacement of warm bodies of air by cooler masses. At night the breezes are reversed, since water cools at a slower rate than land. Similar breezes are generated in valleys and on mountains as warmer air rises along the heated slopes. At night, the cooler, heavier air descends into the valleys. Figure 1-1 illustrates the generation of land and ocean wind formations.

On a broader scale, large circulating streams of air are generated by the more intense heating of the earth's surface near the equator than at the poles. The hot air from the tropical regions rises and moves in the upper atmosphere toward the poles, while cool surface winds from the poles replace the warmer tropical air.

2 FUNDAMENTALS OF WIND ENERGY

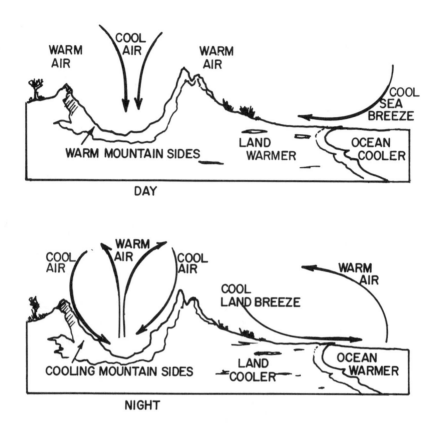

Figure 1-1. The generation of land and ocean winds.

These winds are also affected by the earth's rotation about its own axis and the sun. The moving colder air from the poles tends to twist toward the west because of its own inertia. The warm air from the equator tends to shift toward the east because of its inertia. The result is large counterclockwise circulation of air streams about low-pressure regions in the northern hemisphere and clockwise circulation in the southern hemisphere. The seasonal changes in strength and direction of these winds result from the inclination of the earth's axis of rotation at an angle of 23.5° to the axis of rotation about the sun, causing variations of heat radiating to different areas of the planet. Figure 1-2 illustrates how the earth's rotation affects winds.

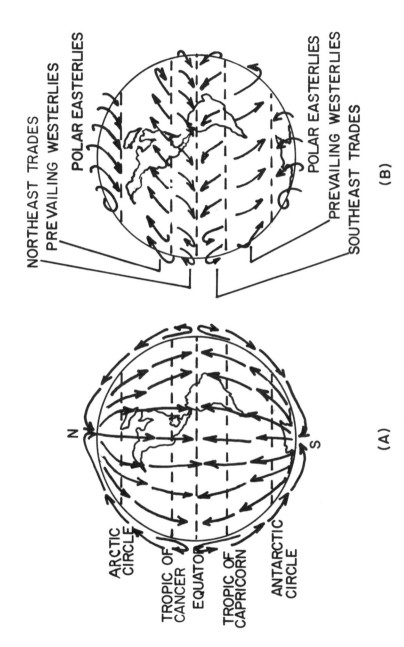

Figure 1-2. A. Direction of winds if the earth did not rotate.
B. Direction of winds affected by the earth's rotation.

4 FUNDAMENTALS OF WIND ENERGY

WIND AS AN ENERGY SOURCE

The use of wind as a source of energy is an old approach dating back some 2000 years to the Persian windmills. Until the birth of the industrial revolution and the advent of the steam engine, windmills ranked second only to wood as an energy source. The traditional applications of wind were primarily as sources of kinetic energy for rural, agricultural and a limited number of industrial uses such as pumping water and grinding grain and feed.

By the 1850s, roughly 25% of nontransportation energy in this country was supplied by the windmill. By the turn of the last century, windmills were widely used for generating electricity. Today, a large number of windmills are utilized for water-pumping operations in western rangelands for watering livestock. Wind-powered electric generators were virtually done away with in the 1930s when centralized electric power became available under cooperative utilities established under the Rural Electrification Administration (REA). By the mid-1940s the use of large-scale wind-operated electric generators was deemed uneconomical, and all major installations were replaced by the more conventional electric power-generating plants. The decline of the importance of wind energy is illustrated in Figure 1-3 by a comparison of wind energy usage to the total energy consumption in the United States.

Currently, U.S. energy consumption is the equivalent of approximately 2500×10^8 kWh per year. A bar graph illustrating the distribution of energy consumption in the U.S. is shown in Figure 1-4. Figure 1-5 illustrates the present sources of U.S. energy. It should be apparent to the reader that our large dependence on oil as a source of energy production must decrease, especially in view of the fact that approximately 14% of our total energy consumption is based on importation. Although the use of oil and natural gas will become less important, the U.S. energy consumption is projected to increase to the equivalent of 6000×10^8 kWh per year by the end of this century unless more stringent conservation programs are instituted.

Because the emergency of the energy crunch is real, a resurrection of wind energy applications has occurred. The amount of energy associated with wind is enormous. Large quantities of energy are constantly being transferred to the winds from the sun.

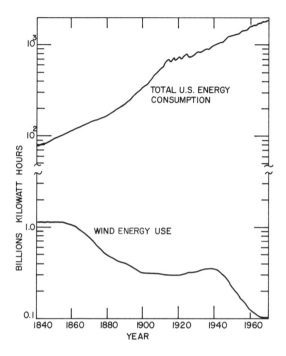

Figure 1-3. A comparison of wind energy usage to total U.S. energy consumption.

It has been estimated that the total power capacity of the winds surrounding the earth is of the order of 10^{11} gigawatts (GW). The 1972 Solar Energy Panel of the National Science Foundation and the National Aeronautics and Space Administration estimated that the power potentially available from surface winds over the continental U.S., the eastern seaboard and the Aleutian Arc is equivalent to approximately 10^5 GW of electricity, which is about 30 times the estimated total U.S. power consumption by 1980. Although these figures illustrate the great potential of wind energy, they do not give the entire picture. The technological problems associated with harnessing any portion of this energy with reasonable economics are substantial. To begin with, winds vary considerably in different geographical areas according to the season and from year to year. Variations in wind direction, speed, strength and temperature have a large effect on the energy. As an example, Table 1-1 illustrates the estimated yearly change in total energy based on a study made at Amarillo, Texas.

6 FUNDAMENTALS OF WIND ENERGY

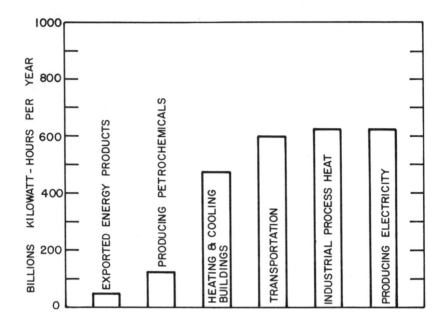

Figure 1-4. Breakdown of energy consumption in the U.S.

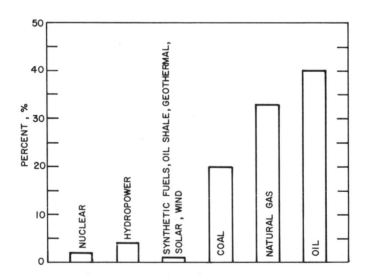

Figure 1-5. Present sources of U.S. energy.

Table 1-1. Yearly Changes in Total Energy of Winds in Amarillo, Texas

Year	Energy (kWh/m^2/yr)	% Deviation From Previous Year
1970	1652	-
1971	2290	38.6
1972	1852	19.1

Further complicating the problem of large yearly and monthly variations, significant differences in winds over limited geographical regions exist. Through the use of contour maps, general observations have been made; however, these generalizations are not enough for wind energy engineering. For the most part, strong winds persist in the polar and temperate regions, while weak winds predominate in the tropics. Winds are usually stronger in oceanic and coastal regions than over the inlands. Furthermore, they are stronger in mountainous areas than on the plains. Unfortunately, these general patterns are often invalidated by large variations. For example, although the plains of the North American midcontinent are relatively windy, the plains of northern India are not. A case study in Honolulu illustrated that the average yearly windspeed over an area with a 10-mile radius varied as much as 65%.

The major problems associated with extracting energy from winds, however, have to do with the physical properties of air. The density of air is small; therefore, equipment designed to remove appreciable quantities of energy from moving air must be capable of intercepting large areas. Thus, the mere size requirements for equipment have limited the practicality of wind energy projects. In general, air is a relatively unstable commodity. Unlike water, air streams cannot be readily concentrated by channeling nor collected to store energy; most important, they change direction, speed and strength with little advance warning.

Although the technical problems are great, they are not insolvable, nor are their solutions far from being economically feasible. Public acceptance of wind energy conversion schemes is necessary in the design of any widespread usage of wind energy. As will be discussed later, the environmental impact of proposed systems is small in comparison to conventional electric power-generating plants. In this respect, wind-powered plants neither contribute to thermal pollution nor discharge chemical effluents as do fossil-fuel or nuclear-based plants, and they have the advantage over hydroelectric systems in that they do not require any flooding of large land areas or major changes to the natural topography.

PAST RESEARCH AND PRESENT INTEREST

Feasibility studies have been initiated in a number of countries to assess the technical and economic problems of wind energy for a number of different applications. The U.S. has undertaken an extensive wind energy conversion system (WECS) program with the primary objective of developing a WECS capable of providing a significant contribution to U.S. energy requirements by the year 2000. To accomplish this, an extensive experimental program is being developed with a short-term goal of establishing several pilot-plant facilities that will be funded jointly by government and public utilities for use by the 1980s. While the federal government's primary interest concerns the use of these systems in electrical energy generation, a number of other areas will be explored, including fuel generation, crop drying and fertilizer manufacturing.

The United Kingdom conducted a study between 1948 and 1962 which made a significant contribution in establishing terminology and developing criteria for site selection. Extensive studies of wind variations and speed were made between 1948 and 1956 across the country and sites were selected. Measurements were made over various time intervals for over 100 of the catalogued sites (all of which were below 2000 ft elevation). Measurements were made with standard-type cup anemometers mounted on simple light-weight guyed structures. Three 100-kW prototypes were constructed and operated. The results of these investigations indicated that the western seaboard and the islands of the British Isles were capable of supplying wind-generated electrical plants with both available and consistent winds. In addition, all three of the prototypes were able to achieve their rated output by either meeting the design-rated wind speed or exceeding it. Major problems, however, were encountered in equipment failure. As will be discussed in a later chapter, these plants and individual units are large, complex systems with substantial maintenance problems.

Hence, a further technical obstruction that must be faced by a WECS is the design of large-scale equipment for long-term usage.

Like the U.S., the U.K. is currently conducting wind-power studies that will affect the ultimate design of large-scale operations. In addition, commercial plans are in the making for a 60-kW wind power plant capable of direct conversion from mechanical to heat energy and coupled with latent heat storage devices.

Since 1973, the Swedish State Power Board and the STU (National Swedish Board for Technical Development) have been involved in feasibility studies aimed at evaluating a WECS. These studies have addressed such issues as:

(a) the development of different wind energy systems and design units,
(b) possible modeling techniques that can be incorporated into design, and
(c) detailed technological and economic assessment of various wind systems and comparison with conventional and proposed alternative energy schemes.

Like the U.K., Sweden has a large number of excellent potential sites for wind energy facilities. Figure 1-6 shows the seasonal variation of total wind energy available in that country. As shown from the plot, the monthly variation in available wind energy follows the electrical consumption demands reasonably well.

Shortly after the end of World War II, France began work on large wind generators that deliver power directly to utility networks. The exploratory work done at that time was short-lived. Several prototypes were examined between 1958 and 1962, but at present no effective research program has been established. Studies by the Electricite de France indicated that France is well endowed with a large number of potential WEC sites. Figure 1-7 indicates approximately the number of sites where the annual wind power could possibly support a plant.

The Canadian government established an interdepartmental group in 1974 to advise the cabinet on appropriate avenues of funding and research topics in all energy programs. In the area of wind energy conversion, attention has been directed largely at high-speed vertical-axis wind turbines.

10 FUNDAMENTALS OF WIND ENERGY

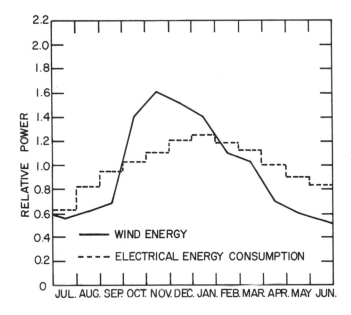

Figure 1-6. A comparison of the seasonal distribution of total wind energy to monthly electrical consumption in Sweden. The ordinate, relative power, is defined as the ratio of the monthly consumption to the total yearly electrical consumption.

Extensive meteorological studies have shown that while there is an excellent supply of wind available for energy production, it is unevenly distributed. Figure 1-8 is a rough indication of the relative wind energy distribution across Canada. Large-scale wind tunnel tests have been conducted on small vertical-axis turbines to test for rotor solidity and rotor drag. Some modeling for large-scale designs has been developed, and small-scale gravity-elastic prototypes have been tested in wind tunnels for gravity sag, blade erection and collapse under high wind conditions.

The Netherlands initiated a WECS centuries ago. Toward the end of the 16th century, a scientist, Simon Stevin, made detailed calculations on windmills and introduced improved designs. Some of this very early work has been incorporated into modern-day designs. Feasibility studies for assessing the potential of wind energy usage in the Netherlands were carried out in the 1920s; however, extensive experimental programs were not initiated until

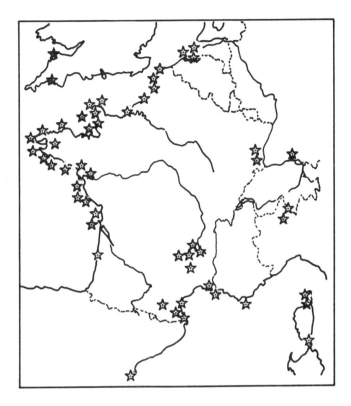

Figure 1-7. Stars indicate the locations of sites where annual wind power in France exceeds 3000 kWh/m^2 (all sites at a 40-m height above ground).

almost two decades later. Studies during the 1950s resulted in a major design, and a 50-kW prototype was constructed. The electric energy production of this system is reported to be between 30,000 and 40,000 kWh/yr. The current Netherlands research program in this field includes detailed meteorological studies on the North Sea as well as analysis of weather data from measurements in the coastal regions. Final design work is being completed on a large-scale vertical-axis rotor, and, as in the U.S., there is considerable interest in exploring the environmental impact that wind-generated electrical plants may have. Figure 1-9 summarizes the major areas of research currently being conducted by various agencies in the Netherlands.

12 FUNDAMENTALS OF WIND ENERGY

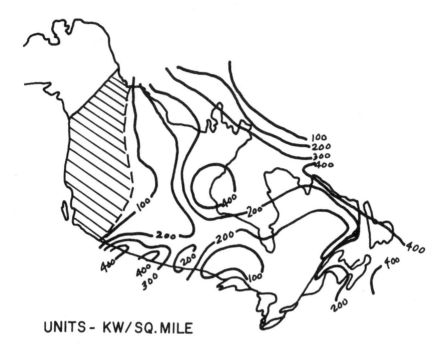

Figure 1-8. Distribution of average wind power in Canada.

The Technion (Israel Institute of Technology) has focused its attention on various designs for shrouds that encase wind turbines. Studies have shown that when these turbines are enclosed in specially designed shrouds, their output power can be increased by as much as a factor of three. These are primarily small-scale studies. Reference to this type of work can be found in the bibliography.

Denmark has probably the longest history of WEC programs. As early as the 19th century, Dr. La Cour initiated systematic studies into the possibilities of wind energy for electricity. During World War II, the F. L. Smidth and Co. manufactured 18 large d.c. WECS with rotors up to 79 ft in diameter for commercial use. Denmark is acutely affected by the energy crisis in that it imports approximately 90% of its energy in the form of oil and the remainder in the form of coal and gas. Furthermore, it is limited in terms of other energy alternatives, such as water power, on which

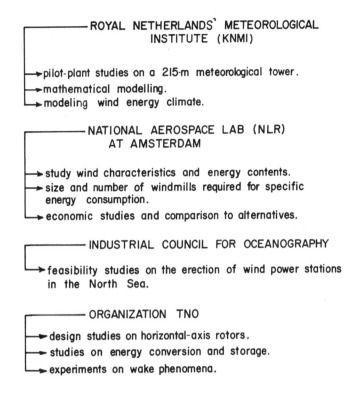

Figure 1-9. Current areas of research by various government-supported agencies in the Netherlands.

Norway and Sweden depend heavily, and exploitation of the Danish sector of the North Sea oil fields has not proven profitable. The Danish government has thus instituted detailed economic evaluations of proposed WEC systems for the production of a.c. electricity that can be fed directly into a utility network. Although this aspect will be discussed at length in a later chapter, it is worthwhile to introduce some terminology at this time.

Breakeven cost is defined as the investment for which the current value of the total costs of a WEC program during the depreciation time (or writing-off period) becomes equal to the present value of the corresponding savings of fuels during that same time span. The depreciation time is valued as the same as the expected physical lifetime of the WECS. It is important to note that this time will vary for the different components in a WECS, such as rotor, tower, and generator. Based on these definitions, then,

14 FUNDAMENTALS OF WIND ENERGY

wind energy usage becomes profitable when a system with specified efficiency, operating and maintenance costs, etc., can be manufactured and operated below the breakeven investment figure. The WECS becomes unprofitable when the investment exceeds this value.

CHAPTER 2

HISTORY OF WIND ENERGY

"Of all the forces of nature, I should think the
wind contains the greatest amount of power."
.... Abraham Lincoln

EARLY DEVELOPMENTS

The earliest known wind machines date back to the ancient Persian windmills circa 200 B.C., which were devised for grinding grain. Basically, these systems were primitive vertical-axis panemones with wind-catching surfaces as much as 5 m in length and up to 9 m tall. Usually the wind-catching surfaces or sails consisted of bundles of reeds. Figure 2-1 shows a sectional view of these early designs.

Windmills were introduced to the western world in the 1100s A.D. The earliest references that appear in the literature are 1105 A.D. in Arles, France, 1180 A. D. in Normandy and 1191 A. D. in England. By the 13th century windmills were used extensively throughout most of Europe.

Near the beginning of the 14th century, the Dutch had become the leading craftsmen in designing windmills. Jan Adriaenszoon is the most noteworthy of the Dutch windmill designers. He is credited with improving the practice of using windmills for removing water from flooded lands.

During this century, Holland was a desolate country. Often tides from the North Sea flooded the barren lowlands and

16　FUNDAMENTALS OF WIND ENERGY

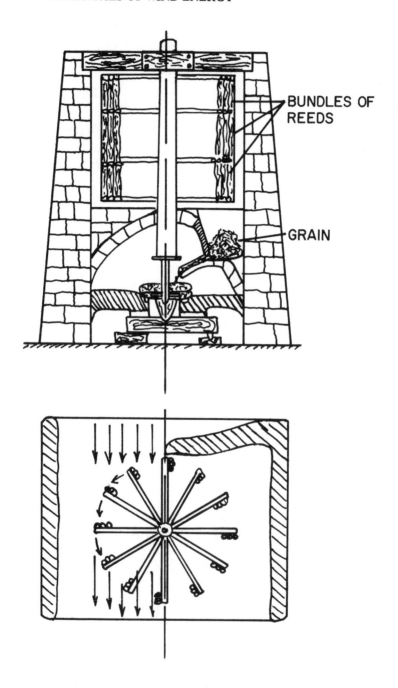

Figure 2-1. Sectional view of early Persian windmill with grindstone.

destroyed communities. Centuries before, considerable effort had been spent in constructing dams and in utilizing windmills for moving water from field to field. Many of these practices were unreliable, however, and numerous large settlements were endangered by water breaking through dikes and dams. Adriaenszoon and others made major contributions in this era to the solution of these problems.

By the middle of the century there were several types of windmills in operation in Holland. Post mills, originally designed to turn millstones, were employed for lifting water. These windmills had a hollow mainpost that accommodated a rotating shaft. The Dutch called these hollow post mills *wipmolens* and used them to turn large wooden wheels with troughs or scoops around them for collecting the water. In addition to this innovation, improvements were made on the wind-catching surfaces. Crude airfoils were invented when it was discovered that the pull of sails on a shaft was greater if the sails were positioned behind rather than centered on the masts or whips that supported them. Boards were then positioned on the other side of the mast to provide a blunter edge, thus making it a fairly effective aircoil.

Another major innovation in windmill design by the Dutch was the tjasker. The tjasker consisted of a large-diameter pipe that housed an Archimedean screw that was turned by the wind-catching sails on its upper end. One end of the pipe was submerged into the flooded field or pond and the sails on the upper end were elevated to capture the winds. Figure 2-2 shows a schematic of the tjasker. This system had the main advantage of mobility; that is, it was considered portable and could readily be moved from site to site.

A third type of windmill that developed during this period of Dutch history and became the forerunner of the 20th century prototypes was the smock mill. This design was so named because its profile resembled an artist's smock. With this system only the cap at the top of the tower had to be turned, rather than the entire tower, to maintain the sails in the proper direction to intercept the wind. This was accomplished either by tailpoles similar to those on post mills or with gears inside the tower. By the 17th century, smock mills were carefully engineered machines used primarily for grinding grain; in addition to being functional, they

18 FUNDAMENTALS OF WIND ENERGY

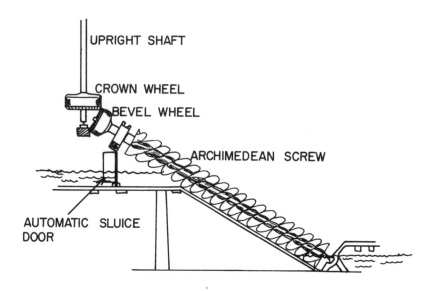

Figure 2-2. Schematic illustrating the Archemedean screw design of a drainage windmill.

were also aesthetically appealing. The large smock grain mills, with their wide high bases, often served as a home, as the machinery was often mounted in a multistoried wooden tower overhead. The sails were generally too high to endanger pedestrians. Usually, these tall structures had "inside winders," which were an arrangement of gears and wheels housed within the tower to direct the sails into the wind. These mills were used to grind chalk, lime, mustard and snuff in addition to grain.

In the late 16th century, the first wind-driven sawmill was developed, opening a new and profitable area of wind use. The first Dutch mills of this type employed wooden cranes to lift logs from the water into long sheds, where they were forced through frames of saws. These units were built around kingposts that were supported by brick pillars in the centers of round brick walls. On the circular outer wall of the foundation an oaken sill, called the winding floor, was located. On the winding floor, the mill could be moved around on rollers to reposition the sails.

Other developments by the Dutch included mills for producing oils and paper. Windmills producing oil crushed seeds in iron cylinders. The pulverized seeds were warmed over hearths and further ground with stamps. Another interesting practice involved mounting a windmill on the deck of a ship for the purpose of pumping water out of the hull. This was common on Norwegian sailing ships during the late 19th century. Prior to the advent of steam and electric power, the use of wind-power was immense. In Holland, it was estimated that there was one windmill serving some function for every two thousand persons. Figure 2-3 shows a Dutch drainage mill, typical of the 18th century.

Another nation that excelled in the design of windmills was England. By the 19th century, it was estimated that there were some 10,000 windmills in operation in the lowlands of the British Isles. The British were among the first windmill builders to replace wooden parts with metal and to use roller and ball bearings. Many of the units were much larger than the Dutch designs. The largest tower mill in England was constructed in 1810 at Yarmouth. This monstrous unit was over 37 m tall with a 12-m round base and walls up to 1 m thick.

In the 1700s, the English experimented extensively on their designs. John Smeaton of Yorkshire, for example, was one of the first builders to introduce cast iron in millwork. This simpified attaching long sails in a rotor. Smeaton did a considerable amount of work in establishing guidelines for designing windmill sails, much of which are still followed today. Some of his early experimental work revealed that the velocity of the sail's tip is almost directly proportional to the wind's speed. He also observed that the maximum load on the sail is almost directly proportional to the square of the wind's velocity, and that the maximum power generated is proportional to the wind velocity cubed.

One of the most significant English contributions in the 18th century was a simple invention by a blacksmith named Edmund Lee. Lee discovered that he could maintain the windmill sails in the direction of the wind by attaching a secondary rotor with several blades to the tower at a 90° angle to the main sails. He observed that when the primary sails squarely faced the wind, the secondary rotor or fantail did not; hence, the blades turned slowly. When the wind changed direction, the fantail whirled faster and was thus able to pull the big wheel to the left or right to keep the wind whirling it. The fantail was in reality one of the first mechanical feedback control devices. By detecting a change in

20 FUNDAMENTALS OF WIND ENERGY

A

B

C

Figure 2-3. A. Typical North-Holland drainage mill.
B. Typical North-Holland corn-mill of the 18th-19th century.
C. Windmills were the subject of both art as well as industry and economics. Their importance is reflected in this 18th-century, 7-in.-high Dutch silver spice box.

wind direction, it was capable of adjusting the direction of the sails much like the ball float on a modern day sump pump can detect liquid level and signal the device to turn on or off. The fantail was immediately accepted and put on many windmills.

Another unique approach to automatic control was introduced in 1772 by Andrew Meikle, who attached wooden slats instead of canvas to the arms of the windmill. The slats were connected to a bar alongside the mast and were held in position by a metal spring. Meikle found that by adjusting the tension on the spring just right, his crude feedback control device would allow the slats to open when the wind's speed increased in order to prevent the sails from whirling too fast. When the wind died down, the slats would close, providing greater windcatching surface area.

At the turn of the 17th century an improved version of Meikle's design was introduced by Sir William Cubitt. In this system the sails were adjusted by bars connected to a "spider" located in the front of the shaft to which the canvas sails were attached. The windshaft was hollow and a rod passed within it to a centrifugal device. In the centrifugal device a weight rose or fell when the wind spun the sails faster or slower, respectively. The action of the weight adjusted the vanes and thus achieved a more stable operation in changing winds. Cubitt's design became quite popular in Denmark and Germany as well as in the British Isles.

Many historians feel that the 18th century was a golden era for windmills in England. Inventions too numerous to mention here developed during this century, and windpower played a significant role in England's cultural as well as industrial development. But the importance of windpower and windmills was short-lived. By the 1800s, the numerous improvements made windmills an almost uneconomical venture to invest in. In addition, with the birth of the industrial revolution came the steam engine. Steam-powered mills with rollers were more efficient and had greater productivity than the old millstones. By the latter part of the 19th century, the major applications of windmills were directed to local farmland requirements in Europe.

AMERICAN WINDMILLS

The early Spanish explorers and settlers introduced windmills to the Americas in the 1500s. The Spanish built many windmills in various colonies; however, very few were actually constructed in this country, mainly because wood was a scarce commodity at many of the mission sites. This is particularly true near the settlements along the Rio Grande and in California. Windmills were constructed in the Danish colonies in the Virgin Islands. The French built windmills in the 1600s along the banks of the St. Lawrence River, later between Lakes Erie and Huron, and in the 1700s near St. Louis and along the delta of the Mississippi River.

The English introduced windmills to the colonies. One of the first was in Virginia and later another was erected near Massachusetts Bay. The Dutch put others near the mouth of the Hudson River. The Germans, Swedes and Portuguese also established their systems in various settlements.

Windmills were encouraged during these colonial days. In 1715 a special decree from the Lords of the Carolinas established incentives for constructing both wind and water mills. For farmers desiring and windmill, for example, the Surveyor General was authorized to section out a half acre of land for its construction. Around the northern region of the Chesapeake Bay both water and windmills were operated. One unit, constructed in 1762 in Camden County, New Jersey, could be operated by either wind or water.

James Edge, a baker, is credited with one of the finest and best designed windmills for his time to provide flour. Built for him by two British millwrights in 1815, it was unique in having metal rather than canvas sails.

During the 1800s windmills moved westward with the pioneers. Some of the early designs were utilized in Illinois, Iowa and San Francisco. The mills in the east were primarily English smock mills. Their frameworks were usually made from oak with pine floorboards and cedar shingles. Along the coastal settlements, windmills were employed as signals for whaling vessels as well as to operate millstones or sawmills. Figure 2-4 shows a typical American windmill during the early 19th century.

As in Europe, windmills were adapted to a variety of uses. The first windmill in Massachusetts was used to grind corn. Toolmakers employed them to make cranberry rakes and shovels. An interesting application was the refinement of salt. In the 1700s salt was scarce and one major source was the sea. Windmills

HISTORY OF WIND ENERGY 23

Figure 2-4. Typical American windmill during the 19th century.

were used to pump salt water from the oceans or estuaries into large wooden vats where chunks of salt were collected after evaporation by the sun. By 1830 there were over 400 solar salt mills in operation in this country. The salt windmills thrived for approximately five decades until deposits were discovered inland and drove the price of salt down.

Prior to the Civil War very few major changes were made on American windmills. However, during the 1860s and early 1870s nearly a thousand new designs and adaptations came about. A variety of fans with many short solid blades were employed, and new and more intricate mechanisms were invented to keep the fans facing squarely into the wind.

During the colonial days, the major application was grinding grain. The emphasis now shifted to the early use of the Dutch; that is, pumping water. The newer systems designed for this purpose were considerably smaller and less attractive than the Dutch or English post and smock mills. Since their sole purpose was to pump water, no accommodations were necessary to house any heavy machinery or millstones. These smaller units had two main advantages: (1) they were considerably less costly than the earlier designs, and (2) they were simpler to construct and required less maintenance. There were three large domestic uses for these systems:

1. Railroads utilized the windmill to fill holding tanks at water stations for their steam operated engines.
2. The upper classes were introduced to the indoor commode and required running water for their bathrooms.
3. The western cattlemen and homesteaders used them for irrigation and watering livestock.

By this time windmill building was no longer a cross between art and science. Americans were treating windmill design as straight engineering and were putting their knowledge on the production line. By 1889 there were 77 factories manufacturing windmills in the United States. By the turn of the century the American windmill was a large part of the export market. Even in Europe, where windmill design had excelled for centuries, the simplified American units were preferred by many farmers. The industry continued to strive for over 40 decades. Figure 2-5 shows the rapid increase in the gross value of windmill parts and machinery on the open market produced by the industry over a period of 50 years. Many of the factories had been located in small trading centers near Chicago. Often these factories were the first and only industry in these towns for many years, employing seasonal farm workers or farm hands that were let go because of the increased mechanization of agriculture.

Two men were prominent in the industry's short-lived growth. One was a Connecticut mechanic named Daniel Halladay, and the other was an Indian missionary in Wisconsin named Lawrence Wheeler.

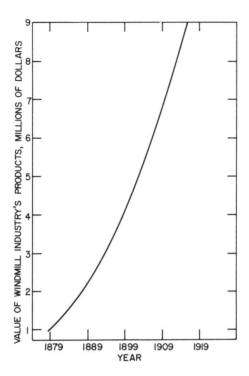

Figure 2-5. The rapid rise in the value of the windmill industry's products.

Halladay is credited with attempting to invent a completely self-regulating windmill. One of the major drawbacks of the early prototypes was that strong gusts of winds or storms could cause considerable damage to the sails. The user often did not have the time to regulate the position of the sails manually or even to provide daily maintenance checks on the system. Windmills underwent considerable weathering in a short period of time simply because of the unpredictability of the wind. Although crude feedback systems had been invented centuries earlier by the Dutch, they were not always maintenance-free. Halladay devised a scheme that regulated the rate at which the wind could spin a vertical wheel around its hub.

The machine he designed consisted of sails constructed from thin wooden planks. These slats were hinged to a ring around the

26 FUNDAMENTALS OF WIND ENERGY

hub rather than to the hub itself as was normally done. They were then connected through a sliding collar located on the shaft to a weight. The weight moved freely up and down to vary the blades' angle. Hence the wind struck the slats' surfaces squarely when it increased or decreased in speed. When the wind speed was slow, the fan wheel was almost flat and the wind made impact with the broad side of each blade in it. When there were strong winds, the slats attached to the ring swung out of the way. Under these conditions the rotor resembled an open-bottom basket through which strong winds could blow without causing damage.

In a later modification, the number of slats or blades was increased, and they were attached to six or eight rods mounted in a circular fashion about the hub. This configuration allowed the use of a dozen or more blades. The design came to be known as a "rosette" because the blades resembled flower petals. As in the previous design, the wheel slats were nearly flat when in full operation, but when strong winds rose and the blades tilted outward to take the wheel "out of sail" it resembled a cylinder. In order to maintain the wooden wheel in the direction of the oncoming a weather vane like rudders was used. Figure 2-6 illustrates the design.

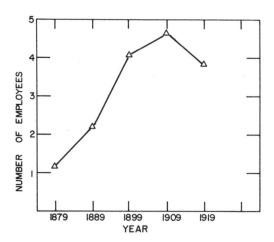

Figure 2-6. Illustrates the growth and decline of the American windmill industry in terms of numbers employed.

Halladay, backed by a man named John Burnham, started a factory in South Covington, Connecticut, in the mid-1850s. Later, because of the growth of the railroad, the plant was moved outside of Chicago and went under the name of the United States Wind Engine and Pump Company. During the decades to follow the company became the world's foremost manufacturer of windmills until its closing in 1929.

The other prominent inventor and industrialist, Leonard Wheeler, patented a wind machine in 1867 that had a solid fan. With the success of this design a thriving family business evolved and presented fierce competition for the Halladay windmills. The Wheeler model was given the name Eclipse windmill. It had a large number of blades directly attached to the hub and the wheel was kept in the wind's path by use of a rudder during mild conditions. A small vane located on the side of the tower adjusted the wheel edgewise into or away from the wind, depending on the magnitude of force imparted on it from the wind. This vane was connected to a weight to that when the wind's force or pressure slackened, the solid rotor would be jerked back into the oncoming wind. The Eclipse's main advantage over Halladay's machine was that it was a simpler mechanism requiring fewer joints. Like Halladay's machine, Wheeler's Eclipses were widely used by the railroads.

In the latter part of the century, a great deal of interest evolved in electrical energy. Unfortunately, the general consensus at the time was that steam-powered generators for producing electricity could never compete with windpower energy, either economically or in quantity. This was still a period of exploration, however, and some scientists did recognize the importance of electricity. Moses Farmer, for example, in the 1860s, showed that wind power could be converted to electrical energy. He demonstrated his idea by using three small fans which drove a magneto that generated sufficient current to light an incandescent bulb. Unfortunately, like many other inventions of that day, it was premature.

The windmill industry continued to flourish in this country and in Europe until the turn of the century. Eventually, the perfection of the steam engine marked the beginning of the industry's rapid decline. Sales of windmills to the railroads, farms and various public institutions dropped alarmingly in the early part of this century. Figure 2-6 illustrates the growth and decline of the numbers of employees within the industry over a period of 50 years. By the year 1919, the industry had been reduced to only 31 factories. The final blow was delivered by REA in the mid-1930s when electrical cooperatives were introduced.

It is important to recognize that windmills did more than supply energy requirements for various applications prior to this century. They had developed into complex, powerful machinery which had a marked effect in transforming rural countries into urban industrial nations. With the emergence of the energy crunch, the wind machine is about to star in a new chapter in the history of man.

CHAPTER 3

MODERN APPLICATIONS OF WIND ENERGY

"Blow wind and crack your cheeks, Rage! Blow!"
. . . . Shakespeare—*King Lear*

AGRICULTURAL USES

In general, the agricultural industry utilizes considerably less energy than the chemical-process, paper and metal-refining industries. Hence, as in the past, agriculture is a strong candidate for wind-power usage. Figure 3-1 compares the energy requirements necessary to produce various farm products with the energy needed for the production of other items of various industries, based on the Btus per dollar value of finished product.

The major sources of fuel for farmland production energy are liquid petroleum, natural and LP gas and electricity; however, the actual farm production level of energy consumption of these is small in comparison to the industry as a whole (see Figure 3-2). The other divisions of the industry that are large energy users include processing, marketing and distribution of products, manufacturing, and family living. Table 3-1 summarizes the fraction of total energy consumption associated with each phase of the industry.

Liquid petroleum is the primary source of energy for marketing and distribution as well as for farm production. In processing, electricity and natural and LP gas are used in quantity. It is important to note that the processing of agricultural products and

30 FUNDAMENTALS OF WIND ENERGY

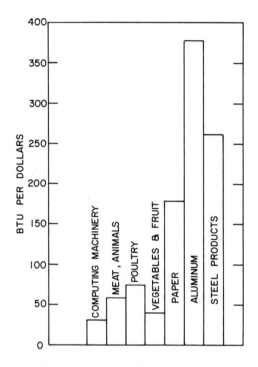

Figure 3-1. Comparison of the input energy requirements of the agricultural industry to other industries.[1]

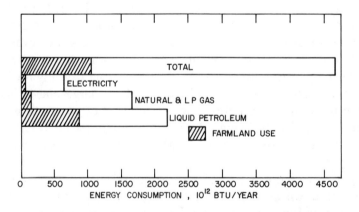

Figure 3-2. Comparison of farmland energy consumption to the rest of the industry.[1]

Table 3-1. Breakdown of Energy Consumption in the Agricultural Industry

	Fraction of Btu/yr Used				
	Liquid Petroleum	Natural & LP Gas	Electricity	Coal	Total
Input Manufacturing	0.008	0.461	0.179	0.060	0.198
Marketing and Distribution	0.381	-	-	-	0.178
Processing	0.039	0.379	0.606	0.777	0.279
Family Living	0.136	0.085	0.135	0.163	0.119

the manufacture of agricultural inputs (fertilizers, pesticides, etc.) consume as much energy as farmland production. In addition, the requirements for marketing and distribution are almost as great as those for production.

In terms of electricity consumption, Figure 3-2 is misleading because it implies that the usage is minimal. In fact, according to 1973 figures, there were nearly 3 million farms in this country, each using on the average 14,400 kWh/yr of electricity. This means that approximately 40×10^9 kWh/yr of electricity is being consumed in the United States for farm production alone and the rate of consumption is continuing to grow. Figure 3-3 illustrates the increasing rate of consumption that farmland production has experienced over a period of 20 years.

As mentioned in the previous chapter, most wind energy usage in the past has been in localized agricultural operations. Today there are approximately 150,000 windmills in the United States, and the majority of them are operating in remote western rangelands. Man has had several centuries of experience with wind energy usage and needs no further research in this area.

Agriculture for the most part has smaller energy requirements than other industries. Hence, any WEC program for farms would require individual units or systems much smaller than would be needed to meet energy demands of other industries. These small systems could conceivably serve as pilot-plant models from which larger units for similar or different applications could be developed.

32 FUNDAMENTALS OF WIND ENERGY

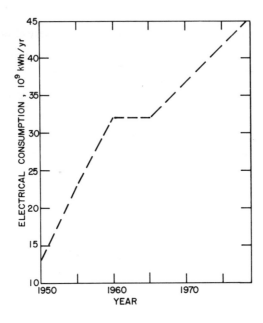

Figure 3-3. Electrical energy consumption for farm production.[2]

One of the major reasons for the decline of wind energy utilization in the early part of this century was the time variability of the wind. As processes and machinery were perfected, production rates increased and far surpassed the windmills. Today's consumers and industry expect to receive goods and serves on demand rather than when they are available. Hence, there is much reluctance on the part of many industries to make use of wind energy unless acceptable energy storage capabilities have first been perfected to supplement the wind's variability.

The agricultural industry is dependent on nature's variables—sun, wind and rain. For the most part, farmers have been able to adapt to these variables by raising different crops in different areas. With a similar form of management it may be possible to utilize the wind's variability to produce energy.

It is essential, if wind power is to be successful here, to recognize and evaluate carefully specific issues, such as the wind's availability, the economics for a particular use or proposed system and the efficiency of the design. In addition, developmental work must be done in the area of energy storage and in perfecting simple energy transmission systems. To some extent this work has been initiated by ERDA (the U.S. Energy Research and Development Administration). Figure 3-4 shows the location of the wind energy program in the ERDA organizational structure. Several areas which are undergoing investigation at present are:

(a) the production of hot water for sanitary purposes. The main application here would be directed toward family living requirements;
(b) refrigeration of buildings or rooms for storage and products;
(c) refrigeration of produce;
(d) drying of agricultural products;
(e) pumping water for irrigational purposes;
(f) heating of buildings;
(g) heating products; and
(h) energy storage for short periods of time, such as tractor and drying operations.

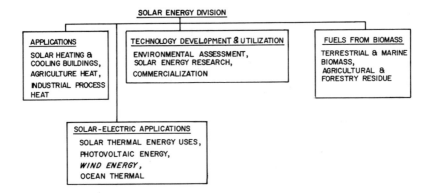

Figure 3-4. Organizational chart of ERDA showing the location of the Wind Energy Program.

Research is currently being conducted in two areas: (1) the conversion of mechanical, shaft horsepower from a wind machine to electrical energy for direct use, and (2) hydrogen as a source of fuel. The Institute of Gas Technology (IGT) is evaluating 5- to 50-kW wind turbines for farm applications. The IGT studies have indicated that the electricity generated from wind turbines can be utilized in conventional a.c. and d.c. applications in generating

34 FUNDAMENTALS OF WIND ENERGY

hydrogen as a combustible fuel. Their proposed system can be used to produce both hydrogen and oxygen that can be stored and later electrolytically recombined to generate power when the wind is insufficient to operate the turbine. Feasibility studies on applications to single-crop, mixed-crop and livestock are being conducted. Farm-oriented energy applications being considered include stationary farm equipment, crop drying, space conditioning and farm vehicles. In addition, studies are also being conducted on household-oriented uses such as water heating, cooking, lighting, operating small appliances, refrigeration and space conditioning. Figure 3-5 shows a schematic of the wind-powered hydrogen/electric system proposed.

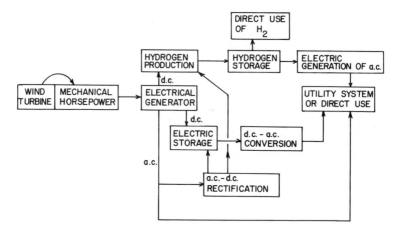

Figure 3-5. Wind-powered hydrogen/electric approach proposed by IGT.[3]

Some of the major factors that will have to be accounted for in the design of an actual system are:

(1) the capacity of the wind-electric generation,
(2) the size of the electrolysis unit required,
(3) the reserve electric-power capacity required for operation, and
(4) the volume of hydrogen and oxygen to be stored plus facilities for long-term storage.

The reserve power considerations may take the form of fuel cells, engine sets, or even conventional batteries. The electricity storage may be in the form of hydrogen in a battery. It is likely that hydrogen would be stored as a compressed gas or a hydride. The major system components are primarily conventional equipment. Figure 3-6 summarizes the primary elements required for such a system.

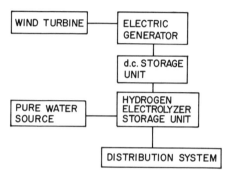

Figure 3-6. Block diagram indicating major components in a wind-powered hydrogen/electric system.

RURAL AND MUNICIPAL USES

Certain limited applications have been suggested for wind power usage in rural and suburban areas. Figure 3-7 shows the major conventional sources of energy used in heating suburban residential areas. As shown, fuel oil and LP gas such as propane are the largest sources. Although electricity is depicted as playing a small role in heating residential areas in rural districts, it should not be considered likely. The rate of electricity consumption is increasing alarmingly, primarily because of the rising cost and potentially limited availability of LP gas and fuel oil.

For the most part, rural areas are widely distributed and often located on sites where wind resources are favorable. For this reason wind energy may have its best opportunity of becoming reestablished as a power supply. As in the agriculture industry, there is a need first to recognize and categorize the different energy uses in rural and remote regions that can possibly be supplied by wind power. In addition, thorough feasibility studies must be made on various kinds of storage systems for specific uses,

36 FUNDAMENTALS OF WIND ENERGY

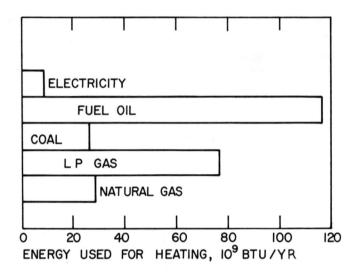

Figure 3-7. Yearly energy consumption for rural residential heating (based on 1970 figures).

with emphasis on the maximum storage capacity required and the rates of charging and discharging of these systems.

Specific applications of wind power usage in rural residential areas are the heating of buildings, generating hot water for sanitation purposes and producing electricity which can be sold to electric utilities. It is feasible that entire wind-driven systems can be developed for the generation of electricity by small rural units. These units could be interfaced with networks of commercial electric power distributors.

Heating needs in rural residential areas might be met by the use of wind furnaces. Wind furnace systems are defined as devices that employ wind and photothermal energy in the production of heat. Studies have revealed that wind energy at many sites is capable of producing sufficient heat on a daily basis; that is, the amount of energy derived from the wind as a function of time over a 24-hr period can in fact match the normal consumer's temperature/day requirements in a home. At many of these wind resource stations, photothermal energy can be readily converted by practical collectors. Solar energy can thus be used to supplement power derived from the wind.

Another application well suited for testing in rural regions is municipal waste treatment by aeration. This can be accomplished by the direct mechanical utilization of wind energy by linkage between a wind turbine and an air compressor. The compressed air can be injected into the bottom of a sewage lagoon, or conventional aerators can be employed. Experimental work on prototypes could be accomplished in small ponds or lakes.

Wind-powered aeration systems can also be used in deterring eutrophication of lakes and ponds and in preventing fish winterkill. Eutrophication is defined as a degradation in water quality caused by an excessive overpopulation of nutrients which produces characteristics that are undesirable for drinking or recreation. Ecologists categorize lakes according to their biological productivity. Nutrient-poor lakes are referred to as oligotrophic. which are usually cold-water mountain lakes with very limited plant growth and low fish production. Mesotrophic lakes, usually greenish in color, have some aquatic plant life and moderate fish production. In eutrophic lakes or ponds, aquatic plant life is in the form of rooted weeds and an abundance of microscopic algae. When this nutrient level is excessive, the aquatic foodchain (Figure 3-8) becomes unbalanced and an overabundance develops of blue-green algae that are not readily consumed by the zooplankton. Under these conditions water becomes turbid and floating masses of algae develop on the water surface where they decompose, producing malodors. This decaying algae often settles to the lake floor and reduces the dissolved oxygen supply. When the dissolved oxygen falls below a certain minimum level, certain species of fish cannot survive.

Most large bodies of water undergo thermal stratification; that is, different temperature zones can be found at various depths in a lake. These thermal zones are subject to changes in season and the circulation rate of the lake. For example, lakes in the temperate zone have two circulations a year (occurring in the spring and autumn). The thermal stratification is direct in the summer and inverse in the winter. During the summer, surface waters warm rapidly, forming a lighter surface layer. This lighter layer is warmer than the lower depths and, as summer progresses, there is a resistance to mixing between the upper and bottom layers which have different densities. The greater the difference in densities,

38 FUNDAMENTALS OF WIND ENERGY

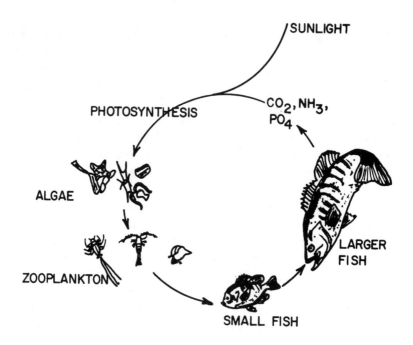

Figure 3-8. The normal aquatic foodchain.

the greater a thermal stratification develops. The upper warm layer is referred to as the epilimnion and is continuously mixed by surface winds and density currents. These conditions are highly favorable for algae growth. The cooler bottom region is referred to as the hypolimnion and is generally dark and stagnant. When eutrophication occurs, the hypolimnion increases in carbon dioxide concentration, resulting in a depletion of dissolved oxygen. Figure 3-9 illustrates thermal stratification during the latter summer months and how it affects eutrophication.

Eutrophication is relatively difficult to deter and is often magnified by outside waste streams that are discharged into the lake. The general approach taken in retarding it is to limit plant nutrients. This usually involves costly treatment methods to nutrient-rich wastewaters. However, in rural areas, sources of nutrient-rich wastewaters that contribute to eutrophication are usually uncontrollable such as farmland drainage, in which the nutrients—phosphorus and nitrogen—are in the form of fertilizers.

MODERN APPLICATIONS OF WIND ENERGY 39

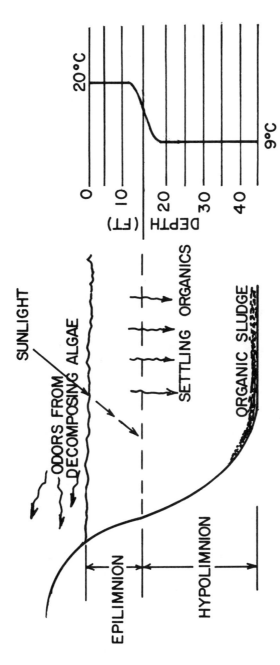

Figure 3-9. Thermal stratification during the summer can cause eutrophication.

Several contemporary methods of control have been employed: artificial mixing, chemical control, flushing, harvesting of algae, aeration and others. Artificial mixing or destratification is accomplished by pumping cold water from the hypolimnion to the surface. This mixing between the warmer and cooler regions causes the dissolved oxygen concentration in the hypolimnion to increase and lowers the temperature of the epilimnion, which retards surface algae growth. It is possible to employ wind turbines in such a pumping operation. In certain cases, where the loss of hypolimnion dissolved oxygen is a more serious problem caused by thermal stratification, deep water diffused aeration can be utilized in place of pumping. This involves employing a perforated pipe on the lake floor to transport and diffuse compressed air to the water. Here, too, wind turbines may be used by linking the wind generator to an air compressor.

During the winter, the warmer or denser water sinks to the bottom, and ice, near $0°C$, forms at the surface. (The maximum density of fresh water is at $4°C$.) This layer of ice acts as a barrier which inhibits photosynthesis, a nonspontaneous reaction by which plant life converts carbon dioxide and water to carbohydrates. Light serves as the energy source for the reaction whereby green plant pigment, chlorophyll, makes use of the energy in the photons in visible light. The reaction provides the basic source of food and fuel. When carbohydrates are metabolized by aquatic life, the reverse of photosynthesis occurs, and the chemical energy that is stored in carbohydrates is coverted to thermal or mechanical energy that is necessary to sustain life. If the lake has a high concentration of organic matter, the dissolved oxygen near the floor gradually diminishes and fish winterkill of certain species will occur. The Colorado Division of Wildlife has conducted extensive studies in mountain lake aeration. Experiments have been performed with electrically powered air compressors that supply Helixors and perforated pipes along the floors of lakes. Air bubbles conveyed to the water either are dissolved or force large holes in the ice, resulting in a replenishment of the dissolved oxygen content and allowing photosyhthesis to take place. A substantial reduction of fish winterkill has been correlated to this method.

MODERN APPLICATIONS OF WIND ENERGY 41

Solutions for the technical problems associated with designing wind-powered aeration systems are not beyond our present knowledge. The major difficulty is in matching the power characteristics of the wind turbine to the load characteristics of the air compressor. The time constant of oxygen consumption in a large body of water such as a lake is often several months; hence, energy storage is not a major consideration. In addition, a lake has the capability of storing large quantities of compressed air as dissolved oxygen during windy periods.

Centrifugal blowers have been considered for matching the wind turbine's load characteristics. Centrifugal blowers produce *pressure* directly proportional to the square of the rotational velocity and *power* proportional to the speed cubed. Recall that John Smeaton in the 1700s noted that the power potential of the wind is also proportional to the wind's speed cubed. However, another design consideration that must be taken into account is that the pressure necessary to inject air at the lake's floor must equal the value of the liquid head to be overcome in addition to increases caused by pressure losses in the supply line. Positive displacement compressors are capable of overcoming the liquid head and are best suited to initiate air flow at low wind speeds which means slower shaft rpm. The major disadvantage of such a unit is that it requires a high starting torque when working against a large liquid head. For this reason, careful selection and design must be made on the actual wind turbine. Multiblade turbines with low- to moderate-speed airfoils may be capable of providing this high starting torque to the compressor.

APPLICATIONS TO PUMPING AND COMPRESSED GAS

Wind power is currently being used in pumping operations for irrigation and watering livestock in this country as it was by the Dutch over 300 years ago. A more modern application of windmill water pumping operations involves pumping water under high pressures to irrigation sprinklers.

Another use for these systems is in pumping water for aqueducts. Large-scale wind machines are employed in providing energy for transporting water from large reservoirs to auxiliary reservoirs in aqueduct systems. This is accomplished by two different approaches: (1) the direct mechanical pumping of water and (2) the generation of electrical power which is used to operate electrical pumps.

Although still in the developmental stage, wind power generators are being tested in pumped-hydro applications. Here, wind systems are employed in the generation of energy used to pump water from auxiliary reservoirs into the main reservoir for a hydroelectric dam. Hence, water being stored in the main reservoir is replenished during windy conditions. This adds to the capacity of the hydroelectric unit in generating base-load electrical energy.

One proposed use for wind energy is in compressing air for a number of different operations, such as in a gas turbine for generating electricity. The electricity produced from these gas turbines can be used for making electricity during peak consumption periods in a public utility system. Proposed systems involve revamping conventional gas turbines by separating the generator, compressor and power stages through clutches. One approach is to operate the motor-generator with a wind machine. The motor-generator would drive the air compressor. Compressed air would then be fed to a large storage vessel. This scheme does away with the power turbine and so no fuel need be consumed. Figure 3-10 shows the compressed air storage system.

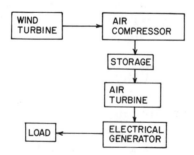

Figure 3-10. Block diagram of the major components in a compressed air storage system.

During the operation, air undergoes adiabatic compression; that is, as the air temperature is raised during compression, there is no loss of heat. A major design consideration in such a system is to provide adiabatic storage by means of a well-insulated storage vessel. If there is little loss of heat during storage, then less heat need be added to the gas when it is finally used in operating a turbine for a specified efficiency. The economics of such a system are favorable in that operational costs would be reduced by using adiabatic storage rather than isothermal because no facilities would be necessary to raise the temperature of the gas up to operating conditions.

LARGE-SCALE ELECTRICITY GENERATION

There have been many attempts in the past to utilize wind energy for electricity generation on a regional scale. Denmark, England, France and a number of other European countries have had relatively large national programs at one time or another. During the 1940s and 1950s, several test systems were constructed and operated. All these units were horizontal-axis, propeller-type prototypes. Most of the programs were abandoned prior to the 1960s because of the availability of inexpensive Arab oil and seemingly favorable projections of abundant, low-cost nuclear fission energy.

During the 1940s private industry in the United States was investing research money in this application. The largest unit in the world to date was designed and constructed (the Smith-Putnam wind turbine) in southern Vermont. The unit consisted of a two-blade propeller with a 53-m diameter swept circle. The propeller was interfaced with a 1250-kWh synchronous generator that was coupled into the local power network. At rated output, the wind generator supplied approximately 10% of the generating capacity of the utility network. The unit was operated as a test system for several years and finally employed as part of the regular generating station. The success of the unit ended when a metallurgical weakness caused a rotor blade to break off. The project was eventually abandoned for many of the same reasons that led the Europeans to discard their programs.

At present, renewed interest has centered on large-scale power-generating windmills that consist of the horizontal-axis, two-bladed propeller type. A prototype has been in operation since

1975 in Ohio and several other units have been designed for experimental studies in utility systems. Proposed designs are expected to have rated outputs up around 1500 kWh with swept-circle diameters exceeding 60 m. A vertical-axis Darrius rotor turbine is planned for the Magdalen Islands in the Gulf of the St. Lawrence River in Canada. This prototype will drive a 200-kWh generator and the connected load in the system, which is currently operated by diesel generators, is around 24,000 kW. Although the proposed system is small compared to the total load, it will serve as a model for large production of wind turbines if successful.

Wind power has also been proposed to generate d.c. electrical power as well as for use with synchronous a.c. electrical generators. The d.c. electricity generated could be used for d.c. appliances or space heaters such as resistance heaters. Another use would be in recharging or storing the energy in batteries which could then be inverted to a.c. loads for later uses. For these types of uses, the energy can also be stored as mechanical motion of a flywheel or in the form of hydrogen and oxygen gases that can be derived from electrolytic dissociation of water. Oxygen and hydrogen can be stored either in liquid form in tanks or as compressed gases in tanks, depleted natural gas wells, etc. Stored hydrogen can be reconverted to electricity via fuel cells or can be burned in gas turbine electrical generators. In addition, the hydrogen can be used as fuel for direct space heating or in industrial process heat.

The major factors that must be considered in designing a system for converting wind energy to electricity are shown in Figure 3-11. It should be emphasized that direct d.c. generation is practical on a small scale at present. Systems currently in use are limited to the range of 10 to 20 kW; the aeroturbine speeds can vary and battery storage systems are usually employed.

For space heating, energy can be stored in a variable-frequency a.c. or d.c. unit in conjunction with a heating coil thermal storage system. Rectifier systems may also be employed to obtain d.c. which can be used for d.c. or inverted to constant-frequency a.c.

For large-scale electricity generation, the energy will be in the form of constant-frequency a.c. The prototypes discussed above employed constant-speed aeroturbines that drove conventional synchronous generators. Constant-speed wind turbines are necessary to operate induction generators which produce the

TYPE OF OUTPUT	WIND TURBINE ROTATIONAL SPEED
d.c. System	Constant Speed
Variable Frequency a.c.	Near Constant Speed
Constant Frequency a.c.	Variable Speed

UTILIZATION OF ELECTRICAL ENERGY OUTPUT

Battery Storage
Interconnection with a.c. Grid
Other

Figure 3-11. Basic factors and alternatives that must be considered in the design of a wind-powered electrical generating system.

constant-frequency a.c. power that can be pumped to utility grids. Suggested reading on synchronous and induction generators for WEC systems can be found in the reference section of this book.

A recently proposed alternative to constant-speed wind turbines is to permit the turbine speed to vary with the wind. This involves employing variable-speed constant-frequency generating systems to achieve constant-frequency power which can be transmitted to existing utility lines. There are two approaches to obtaining constant-frequency a.c. electrical power from variable-speed shafts, namely, differential and nondifferential methods.

In general, the differential method utilizes mechanical techniques to obtain constant speeds. Synchronous generators are employed. The major components of such a system consist of variable ratio gears, planetary gear systems, and a hydraulic pump-motor arrangement. To get constant frequency, a frequency makeup generator that achieves differential action by feeding slip frequency power to the rotor, is used. Non-differential methods employ static frequency changers (a.c.-d.c.-a.c. linkage). The rotary devices are complicated schemes. Some of the proposed rotary devices include a.c. commutator generators, cycloconverters and frequency changers, a.c.-d.c. converters (square wave modulation), and field modulation and demodulation arrangements (for high frequency and low frequency switching).[4,5]

Preliminary work on field modulated frequency systems has shown them to be economically favorable. Essentially, they are specially designed three-phase high-frequency alternators that can be employed in conjunction with power-operated solid-state electronic circuitry which is capable of obtaining single-phase output at low frequencies. A rotating field coil is excited by an a.c. signal at low frequencies.

The early prototypes between 1940 and 1960 were operated with synchronous generators and induction generators; both have severe disadvantages. When the wind turbine unit provides a small contribution to the load, the machine will operate in synchronism with the utility system for variations in wind energy inputs. However, during gusts there is a large tendency for the turbine to run out of synchronism, causing large instabilities in the utility system if the WECS is a large contribution.

When induction machines are operated above synchronism, they behave like generators. Induction generators will continue to supply energy back to the system at the system's frequency provided it is operating above synchronism. They are, however, much less efficient in energy production than the synchronous systems and tend to function at a low power factor resulting from heavy magnetization currents. These magnetization currents are almost in quadrature with voltage. Therefore, present emphasis is on the development of a.c. commutator generators and rotor-fed induction generators.

Although there are a number of alternative approaches to incorporating a WES in a public utility, and many of the technical problems can be worked out, wind energy utilization may be impractical if the economics of such an undertaking cannot be justified. Cost feasibility studies must be made based on our present and future energy needs. Extensive information is required on all the various alternative wind energy systems, along with expected growth rates, demands, fuel, equipment and other aspects such as further process development and research, all of which play an important part in the economic profile. How each of these factors interacts with the others must be determined from projections of the kWh available from wind generating systems of various ratings. These in turn will be dependent upon wind statistics and characteristics.

The economics involved depend on many parameters, including costs associated with equipment, fuel, transmission system, insurance, maintenance and interchange-power arrangements. They also depend upon hourly, daily and annual energy production of a WES. Evaluation of these factors requires detailed information and correlation of kWh short-term and long-term demands, reactive KVA capability requirements and storage capabilities.

Detailed analysis must be made of a planned investment by evaluating the effect that such an investment will have on the company's income. Cash flow, earnings, income tax, assets, preferred stock dividends, etc., will have to be accounted for in evaluating WEC programs.

CHAPTER 4

WIND MACHINES AND GENERATORS

"Who walketh upon the wings of the wind."
.... *Psalms* 104:3

WIND CHARACTERISTICS AND TERMINOLOGY

The energy associated with the wind is in the form of kinetic energy. The kinetic energy (E) per unit volume of moving air is expressed as

$$E = \tfrac{1}{2} \rho u^2 \qquad (4.1)$$

where u is the average linear wind velocity and ρ is the density of the air. The wind speed (or velocity) is proportional to the number of unit volumes arriving at the blades of the wind machine per unit of time. For example, a 15 m/sec wind delivers 15 m³/sec of air to a turbine that has a cross-sectional area of 1 m²; *i.e.,* uA is the number of volume units of air per unit time where A is the cross-sectional area covered by the blades perpendicular to the wind direction.

For a freely moving stream of air with cross-sectional area, A, the amount of power associated with it is equal to the product of this area times the velocity of the wind stream times its kinetic energy per unit volume. Wind power density, P, is defined as the power per unit cross-sectional area of the windstream and is given by

$$P = \tfrac{1}{2} \rho u^3 \qquad (4.2)$$

50 FUNDAMENTALS OF WIND ENERGY

The total available power in the wind will increase with the subtended area of the wind stream. For example, an 8.9-m/sec wind with a 9.3-m cross-sectional area of interest will contain approximately 4 kW of power at sea level, whereas a stream with a cross-sectional area of 9.3 x 10^4 m² will have nearly 40 MW. Figure 4-1 illustrates how the power in the wind stream is affected by the cross-sectional area of flow.

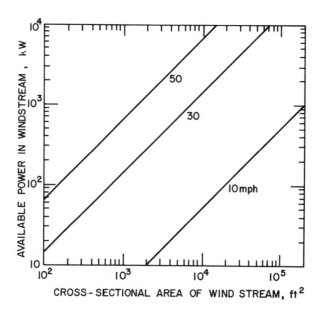

Figure 4-1. The effect of cross-sectional area on the total power associated with a wind stream.

Wind characteristics can vary greatly from one geographical location to another. In addition, the wind at a given site may show large daily or even hourly variations in direction and speed. Variations in the wind speed result in large changes in the power densities. For example, the power density for a wind speed of 4.5 m/sec would have a power density of 54 W/m², but if the winds gust to 13 m/sec, Equation 4.2 shows that the power density increases to 1507 W/m². Average velocities are also known to undergo significant changes seasonally.

WIND MACHINES AND GENERATORS 51

This type of meteorological data is statistically analyzed and is generally recorded in graphical form. In this manner average variations in wind speed or wind power can be evaluated in selecting a wind energy site. One type of plot that the wind engineer finds useful is called the annual average velocity duration graph (Figure 4-2). Such a plot shows the number of hours per year that the wind speed achieves a specified hourly mean value at a given site. Another useful representation of wind power variations is the annual average power density distribution plot, which is a collection of curves depicting the variations of annual average wind power per unit of subtended area.

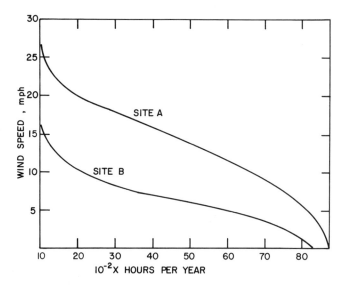

Figure 4-2. Typical annual average velocity duration curves for different geographical sites.

Wind machines cannot be designed on the basis of average wind velocity alone. Detailed statistical information such as the standard deviations in velocity or power associated with the wind must be carefully weighed against any design based on mean or average wind speeds. This type of information must also be heavily considered for daily and even hourly changes in speed, although this is

52 FUNDAMENTALS OF WIND ENERGY

more difficult to analyze because of the wind's unpredictability. Wind gusts, for example, are seldom steady, and there are relaxation periods where the winds are calm. Figure 4-3 illustrates the various information that can be obtained by monitoring the instantaneous wind speed. The instantaneous variations in wind speed resemble the formation of surface waves on a lake or a fluid flowing down an inclined plane. However, wind gusts, unlike waves, do not always show general trends. There may be no detectable frequency for a given gust amplitude at a specific site (frequency being the number of gusts of a specified amplitude observed per unit time). In addition, relaxation periods and their amplitudes may show no detectable trend. Therefore, mean or average values may not reflect a reasonable estimate of the actual wind power available at a given site.

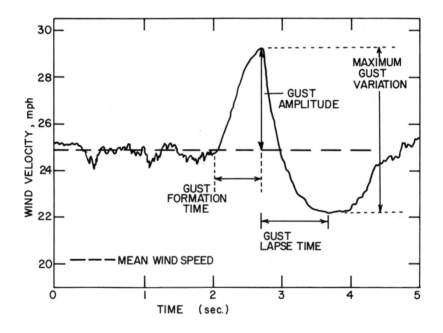

Figure 4-3. Variations in the instantaneous wind velocity show that gusting and relaxation periods can distort average wind speed values.

The most common device used for measuring wind speed is the vertical-axis rotating-cup anemometer. Other devices that are employed for wind speed measurements are the sonic anemometer, the hot-wire anemometer and the pressure-plate anemometer (transducer). Up to and including the 1930s, most instruments employed were referred to as totalizing types, in which the passage of a number of kilometers of wind past the instrument in a given time interval could be converted to kilometers per hour or miles per hour. These instruments were not capable of measuring wind gusts or instantaneous velocities. During the 1940s, however, aviation interests prompted the development of the magneto, which allowed near instantaneous wind velocity measurements.

A wind gust is defined in meteorological terms as a speed which exceeds the lowest observable velocity by a minimum of 10 knots over a specified period of time. Wind velocity data are often reported by frequency or percentage frequency, whereby the occurrence of wind in various speed classes by direction is noted. These speed classes may vary considerably depending on the geographical location of the testing site and the height at which measurements were taken.

Crude estimates can be made of the median wind velocity and of wind duration profiles, $i.e.$, the cumulative frequency distribution of wind speeds. However, more exact estimates and analysis techniques must be perfected in order to obtain better wind profile formulations that can be incorporated into wind machine design. Preliminary studies in this field have indicated that there is an optimum height that must be considered in constructing an aeroturbine. This optimum height is dependent upon the mean wind velocity and the shape of the wind power duration curve which in turn are affected by the physical and climatological characteristics of the site. Meteorological studies in Sweden[6,7] have revealed that the wind power duration profile can be well correlated by empirical formulations:

$$\frac{t(u)}{8760} = \phi = e^{-ku^{\alpha}} \qquad (4.3)$$

where t(u) is the number of hours in a 365-day year in which the wind exceeds a specified velocity, u (m/sec), the constants k and α are determined from average wind speed data by means of a least-squares method from actual wind duration curve and 8760 is the number of hours in a year.

This type of information must also be incorporated with wind gust characteristics and relaxation periods, although it is not exactly clear at this time how this should be done. Gustiness characteristics are a topic of considerable investigation. Available data on peak wind speeds and gust duration times must be assessed.

Historical meteorological data are simply not sufficient as a basis for wind machine design. Many of the standard techniques are outdated or do not provide a complete picture of turbulence structure of the air near the earth's surface. Information on the turbulence energy spectra, particularly the longitudinal wind spectra, is necessary. Studies have shown that measurements made over different topographical regions indicate that the characteristic scale of wind spectra varies significantly. That is, roughness elements such as forests, hills, valleys and structures determine the nature of the turbulence structure in the wind. For example, the turbulence structure over a flat plain or some homogeneous terrain is very much different from the structure over an area that is scattered with houses or a few patches of trees.

In the United States, the National Climatic Center (NCC) is responsible for determining the availability of meteorological data required for the design and site selection criteria for wind-powered machinery. The NCC is part of the National Oceanic and Atmospheric Administration/Environmental Data Service, whose responsibility it is to collect, process and disseminate meteorological data within the continental United States. Ongoing surveys are being conducted by the NCC, which summarizes the massive collection of land and ocean surface wind measurements, upper atmospheric wind measurements, turbulence characteristics, inversions and gust measurements that have been taken beginning in the latter part of the 19th century. Both Sweden and Denmark have similar national programs for assessing wind characteristics.

The statistical data gathered by the NCC and other agencies can be transformed into useful design information for the wind engineer. For example, an annual distribution of the wind power (as described above) can be obtained. The annual average wind energy density distribution is given by the expression:

$$\bar{E} = \frac{\rho u^3}{2} t \qquad (4.4)$$

where $\bar{E}$ is the average wind energy density distribution per unit area (kWh/m²-yr) and t is the number of hours that the corresponding average wind speed occurs during a year.

Plots prepared from the distribution of the annual average energy density for various wind speeds at a given site can be used in determining the speeds at which the majority of the wind energy exists. Figure 4-4 indicates such a plot. It is of interest to note that most of this energy exists above the observed mean wind speed. Furthermore, the contribution to the total annual mean energy content of all velocities is generally small for winds having speeds in excess of approximately three times the average wind speed.

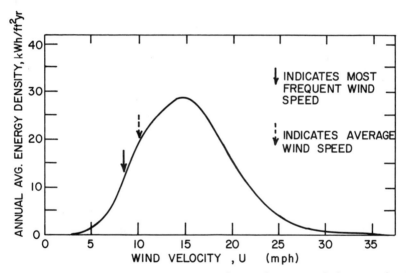

Figure 4-4. Plot shows the distribution of annual average wind energy densities for various wind velocities.

One factor which makes it extremely difficult to interpret average wind characteristics is that velocities tend to vary with height. The NCC, for example, reports average wind speeds taken with anemometers at heights ranging anywhere from 6 to 60 m above the ground, and often the height at which measurements were taken is not reported. Considerable effort has been made

56 FUNDAMENTALS OF WIND ENERGY

recently to standardize anemometer heights at weather stations across the country.

In general, the complex nature of the atmospheric boundary layer, and the flows of heat and momentum in the atmosphere, are poorly understood. There are no well defined theories or models that adequately describe the nature of turbulence in and above the boundary layer. In order to achieve proper design of wind generators, a thorough understanding of the variation of wind velocity with the height is necessary. The shape of the wind profile as a function of height is dependent upon the geography, topography, surface roughness factors and atmospheric stability of a specific site. The profiles may also depend on the wind direction.

Free air that flows above the boundary layer travels at nearly twice the speed of wind over flat terrain. In general, the wind speed increases with the one-seventh power of height. Hence, wind power increases to the three-seventh power of height. This general rule of thumb has been observed to be valid in the range of wind speeds capable of supporting moderate-size wind generators. Justus[8] made a more systematic study of this problem by evaluating wind speed data from four different wind turbines. Weibull scale factors were defined at each site, where

$$p(u)du = (k/c)(u/c)^{k-1} \exp[-(u/c)^k] \, du \qquad (4.5)$$

Equation 4.5 is the Weibull distribution function, where p(u) is the output power of the turbine as a function of speed (u), parameter c is related to the mean wind velocity (units of speed) and k is a variance factor that is inversely proportional to the variances of the speeds about the mean velocity.

Justus observed that the scale factor, c, varies with height as a power law:

$$c/co = (h/h_o)^n \qquad (4.6)$$

where h is the height (meters) and co and h_o are known base cases. The exponent, n, was found to have a value of 0.23 for all the sites tested.

Other correlations have been developed for predicting wind speed as a function of height[9,10]; however, they were largely developed at sites having homogeneous topography. Very little information is available on the speed-height profile over rough terrain such as mountains, valleys or highly populated areas.

Another parameter which is difficult to evaluate on a quantitative scale is the wind's variability. As pointed out earlier, wind varies seasonally, synoptically and over short-term intervals. The highest content of power can be extracted from strong winds. However, in most areas these are not consistent and in addition require relatively strong structures to harness their power. Therefore, a properly designed system must have a cut-off speed arrangement. On the other hand, weak winds with relatively low speeds, although they occur more frequently, contain little power and require highly efficient wind energy collectors. Hence, cut-in speeds must also be incorporated into the ultimate design.

Wind machines are designed to operate within a certain wind speed range and efficiency. It is not practical to extract all the power from the wind over a given time interval, not only because a device designed to do this would have to have exceptionally fast response and be highly sensitive to wind variability and direction, but also because the air downstream of such an aeroarogenerator would have zero velocity and air would thus accumulate. One general class of wind machines are the horizontal-axis turbines. These machines operate within an upper boundary called the Betz limit. The theoretical Betz limit is the maximum fraction or percentage of the power in the wind which can be harnessed; the Betz upper limit is 59.3%. Actual horizontal-axis turbines having two or three propellers are capable of extracting up to 60-80% of the wind's energy but only under ideal conditions. The Dutch multivane machines are capable of considerably less.

The first windmills were vertical-axis machines. These devices are not subject to the Betz limit; however, they do have an upper limit which varies with the specific type of vertical-axis arrangement.

The efficiency, or the degree of power extraction from the wind, is governed not only by aerodynamic considerations, but also by the specific application to which it is adapted. For example, if the wind turbine is to be used in generating electricity, the machine must be matched to meet power ratings in the electrical generator. As an example, if a wind machine were designed to operate at an average wind speed of 8 m/hr (the speed at which a wind machine is designed to drive the generator to its full rated power is called the rated speed), to drive an electrical generator rated at 10 kW, and if gusts up to 10 or 11 m/hr rose (assuming our turbine has no arrangement for high-speed cut-off), then the turbine would deliver increased power to the generator, exceeding its rating and consequently burning it out. Hence, to avoid the problem of supplying more mechanical power than the electrical generator is capable of handling, certain additional design constraints must be made on the efficiency of the turbine.

Gusting is not the only condition limiting the turbine's efficiency. If, for example, the winds undergo a relaxation period, then the power in the wind can be reduced significantly. For the purposes of this discussion, assume that the wind velocity drops down to 6 m/hr. From Equation 4-2 we observe that the total power rating of the wind is reduced to $(6/8)^3$ or only 42% of the power rating in which the turbine was designed to operate. Hence, the electrical output from the generator will be considerably less than 10 kW. When the turbine is operated below its rated speed, the wind generator is delivering less than its designed output; however, when the winds are above the rated speed, the turbine is generally not designed to operate above the rated output but only at the design specification.

TYPES OF WIND MACHINES

Horizontal-Axis Rotors

Many different types of wind machines were designed and constructed during the course of windmill history. Several were discussed in Chapter 2. Today, the most common wind-powered machines are classified according to their axis of rotation relative to the direction of the wind. The major categories of wind machines are:

- (a) horizontal-axis
- (b) cross-wind horizontal-axis rotors,
- (c) vertical-axis rotors,
- (d) thermoelectric type machines,
- (e) translational wind machines.

The last two have relatively little application at this time and are still in the initial development states, and therefore will not be discussed.

Horizontal-axis rotors, often called head-on machines, are those devices in which the axis of rotation is parallel to the direction of the wind. These designs fall into the conventional type of windmills. These devices can either be lift or drag systems.

Lift devices are preferred because they generally develop much higher force per unit area than direct drag devices. Furthermore, drag devices are usually not capable of moving faster than the wind velocity. Lifting surfaces, on the other hand, are capable of attaining high tip-to-wind speeds. This means higher power-output-to-weight ratio which in turn favors the economics of such a system with a lower cost-to-power output ratio.

Designs have been varied with the number of blades. Systems can range from one-bladed units equipped with a counterweight to multiblade systems having 50 or more blades (such as the U.S. multiblade windmill). The bending loads on the roots of the blades are often reduced by employing canted blades. Most horizontal-axis rotors are yaw-active, which means they will change position depending on the wind direction. For small units designed to yaw, a tail-vane is used. For larger systems, the yaw action is generally servo-operated. Figure 4-5 shows several blade configurations for horizontal-axis rotors.

Designs may vary considerably in this class of wind turbines. Blades are usually coupled directly to the output of the system through a shaft on which the rotor sits. Other arrangements utilize a circular rim that is attached to the bladetips. This rim then drives a secondary shaft that is mechanically attached to the power output network.

One class of horizontal-axis rotors is designed so that the blades rotate in front of the tower with respect to the wind direction. These are generally referred to as upwind rotors. Another variation, called downwind rotors, has blades rotating in back of the tower. Figure 4-6 shows both types. They are usually equipped with flaps on the blades and pivot vanes mounted parallel to the blades to prevent the propeller from overspeeding in high winds or gusts and to manipulate the propeller sideways to the wind, respectively. A variety of other techniques are utilized

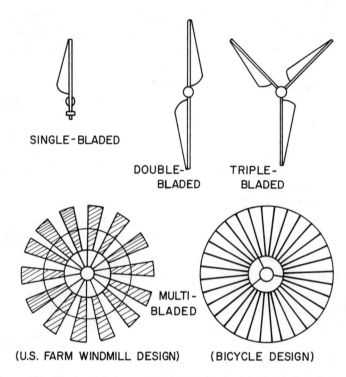

Figure 4-5. Typical blade configurations for horizontal-axis rotors.

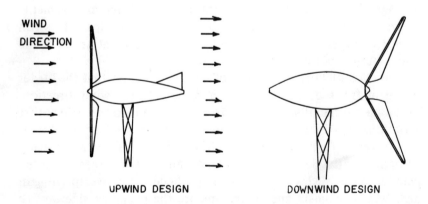

Figure 4-6. Upwind and downwind horizontal-axis rotor designs.

in preventing the propeller from overspeeding, such as feathering the blades or using flaps that rotate along with the blades.

A number of different systems have been developed with counter-rotating blades, while other proposed designs include multiple rotors on a single tower which would tend to reduce tower costs for a specified power output for the system. An area of interest which has the potential of increasing wind machine performance is the effect of shrouds. Tapered shrouds have been suggested for use on concentrating and/or diffusing the wind as it passes through a horizontal-axis turbine. This has the effect of increasing the air stream's velocity and reducing the turbulent nature of the wind. Other specially designed shrouds have been suggested for generating more turbulence by producing vortices around the turbine to concentrate air streams, resulting in an increase in the propeller's angular velocity. Extensive experimental studies have been carried out over the past 15 years on shroud design. Many of these are cited in the bibliography.[11-14]

The major advantages that can be derived from utilizing shrouds are:

(a) The axial velocity of the turbine portion tends to increase for a steady upstream wind speed. This makes it possible to build smaller rotors which can operate at higher rpms (rotations per minute).
(b) The shroud is actually an enclosure around the blades and hence can greatly reduce tip losses.
(c) The turbine axis would not have to be rotated in a direction parallel to the wind for small changes in wind stream direction.

Shrouds are normally designed with a diffuser portion. A diffuser-type shroud is shown in Figure 4-7A. Its purpose is to reduce turbulence, but, more importantly, it imposes a positive pressure gradient on the low-pressure air stream which leaves the turbine, causing the turbine exit pressure to increase toward the atmospheric level before being discharged from the shroud. Figure 4-7B shows a concentrator which can be used to induce vortices around the turbine and increase the air stream's velocity, which causes an increase in the propeller's angular velocity.

One relatively new modification to horizontal-axis rotor designs is the sailwing windmill, developed at Princeton University.[15]

62 FUNDAMENTALS OF WIND ENERGY

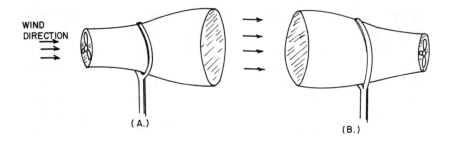

Figure 4-7. (A) Diffuser-type shroud, and (B) concentrator.

Sailwings have the same general shape of solid propellers but consist of a rigid tubular leading edge which has short bars attached to it forming a rigid tip and root. A stretched cable, which serves as a trailing edge for the blade, extends between the free ends of the tip and the root bars. The blade's surface generally consists of a dacron envelope which slips over the tubular leading edge and cable. The envelope is stretched by tightening the cable. This design has the major advantage over its contemporaries in that it is extremely light-weight and relatively inexpensive. The major disadvantage of such a unit is that it is limited to small-scale wind machines because of its materials of construction.

Vertical-Axis Rotor Systems

The first windmills were of the vertical-axis type; however, they were never incorporated in any large-scale operations like the horizontal-axis systems. The one major advantage that these systems have over the horizontal-axis units is that they do not have to be repositioned into the direction of oncoming wind as the wind stream direction changes. This advantage essentially reduces some of the complexity in design and at the same time reduces the gyro forces on the rotors that in the former systems cause stress on blades and other components when yawing.

In general, vertical-axis rotors are those machines which have their axis of rotation at right angles to both the earth's surface and the direction of the wind. Savonius rotors, which were invented and used in the 1920s and 1930s in Finland, are the most common design. The Savonius-type rotors generally employed S-shaped blades—a minimum of 2 curved blades in the shape of an S; up-to-date designs were multibladed units. The Savonius models, like most of the various types of vertical-axis sytems, are primarily drag devices, although this design does actually provide some lift force. The main disadvantage of such arrangements is that they have relatively high starting torques in comparsion to lift-type designs. Furthermore, they have a lower tip-to-wind speed ratio than the horizontal-axis lift systems, which means lower power outputs for a given rotor size and weight.

A major problem associated with vertical-axis systems in general is that they must return captured portions of air streams against the wind. Figure 4-8 illustrates this problem, for the S-shaped blades commonly employed. This disadvantage, however, has been overcome largely by shielding the discharged air with a windscreen or by articulating the discharged stream in a modern idiom such that the blades present a large surface to the wind when traveling downwind and a small surface when traveling upwind. Another approach is to employ an asymmetrical shape, as in the common cup anemometer.

With the exception of those few special designs briefly mentioned above, blades undergo rotation because of a differential gradient in drag forces between the downwind-moving and upwind-moving air streams. The machines governed by this effect are limited in rotational speed by the fact that the downwind-moving air elements cannot travel faster than the wind. In general, such systems are not capable of extracting large quantities of energy from the wind. In addition, these units must utilize a cross-section of material which is greater than the cross-sectional area of the wind intercepted. In such cases, the amount of materials of construction can offset the economic advantage of being less costly because of simplicity in design.

The most promising competitor to the horizontal-axis designs is the Darrieus-type rotor. It was first introduced in France by G. J. M. Darrieus in the 1920s. Its configuration consists of only two or three thin blades that rotate very rapidly. The speed of rotation of the Darrieus is primarily dependent on the machine's overall diameter. The interesting characteristic of this design is that the outermost parts of the blades revolve at three to four times the wind velocity. The blades on the model have airfoil

64 FUNDAMENTALS OF WIND ENERGY

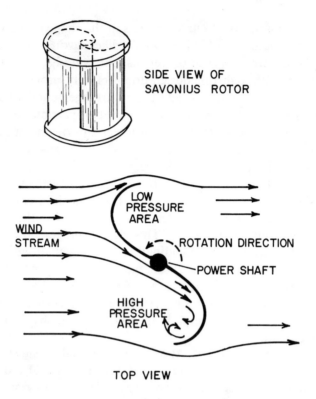

Figure 4-8. Illustrated that the Savonius S-shaped design has the disadvantage of returning wind-catching elements against the wind.

cross-sections. The Darrieus has relatively low starting torque but a high tip-to-wing speed ratio. Hence, these units can be almost as efficient as a horizontal-axis propeller arrangement. A variety of Darrieus rotor configurations have been proposed (ϕ-Darrieus, $\triangle$-Darrieus, and γ-Darrieus). Two different Darrieus rotor arrangements are shown in Figure 4-9.

The Savonius rotor has been suggested as the starting prototype for scaling up and designing an efficient Darrieus. Extensive experimental work and mathematical modeling is being done on Savonius rotors.

The National Research Council (NRC) of Canada is currently involved in a program aimed at developing the Darrieus prototype into a commercial unit for electrical energy generation. The NRC

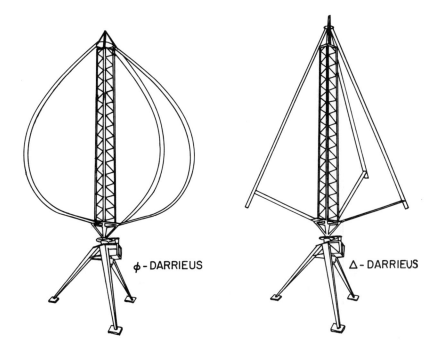

Figure 4-9. Two different designs of the Darrieus-type rotor.

considers the Darrieus the most promising of all the wind machines available for electrical power generation because of the following:

- (a) Due to their vertical symmetry, the need for yaw control is eliminated. As previously noted, this reduces some of the design considerations that would normally have to be accounted for in horizontal systems.
- (b) These units are capable of delivering mechanical power at ground level. Gearboxes and generators can be mounted on the ground and not at the top of the tower as with horizontal-axis wind turbines.
- (c) The Darrieus requires a simple tower support which reduces construction costs and design considerations.
- (d) Because of the simplicity of the blade design and because they are relatively thin, blade fabrication costs are reduced.
- (e) They require no pitch control for synchronous applications. In horizontal-axis wind machines, the pitch control is incorporated in the design criteria for regulating power generation

at a rated generator power, which is dependent on the rated wind velocity of the wind turbine. Figure 4-10 illustrates how power regulation can adversely affect the performance on a horizontal propeller-type turbine. Based on theoretical considerations, however, the Darrieus design will generate a self-regulating power-wind velocity response relationship when operated with a constant rotational speed. This results in power regulation without the need for pitch control.

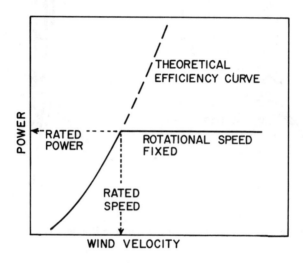

Figure 4-10. The effect of power regulation on the power-wind response curve (performance curve).

It is worth noting that the Darrieus has great flexibility in design. A number of different configurations that are a combination of the Darrieus and Savonius and/or other models have been proposed.

Magnus rotors of the Madaras and Flettner are another class of vertical-axis rotors. Essentially these systems consist of revolving cylinders. These systems first appeared in the United States in the 1930s. When a cylinder is subjected to a fluid stream flowing about its vertical axis, translational forces are generated at right angles to the fluid motion. Known as the Magnus effect, this is illustrated by Figure 4-11. Note that large vortices or turbulence are generated in the fluid stream as it passes the cylinder. Although

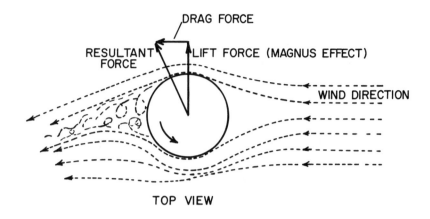

Figure 4-11. Illustrates the Magnus effect on a rotating cylinder.

interesting, the system has been considered more of a curiosity than one having any practical merit. A possible application of this type of rotor is in propelling ships. One rather unique scheme proposed for generating electricity with this device is the Madaras concept. In this arrangement, the spinning cylinder sits on a tracked carriage. The motion of the spinning cylinder causes the carriage to move over a circular track and the carriage wheels to drive an electrical generator.

Cross-Wind Horizontal-Axis Rotors

These devices are similar to water wheels in appearance. They are machines whose axis of rotation is horizontal with respect to the ground and at a right angle to the direction of the wind. The most common type of design is the cross-wind paddle rotor shown in Figure 4-12A. There are a variety of prototypes that have been proposed and constructed. Combinations of horizontal cross-wind rotors and vertical wind turbines such as the "cross-wind Savonius" have been suggested (Figure 4-12B). In general, however, these units are not considered promising from the standpoint of applicability. They are primarily complicated designs that have no marked advantages over either the conventional horizontal-axis turbines or the vertical-axis systems. They must be repositioned into the wind stream just as the standard head-on

68 FUNDAMENTALS OF WIND ENERGY

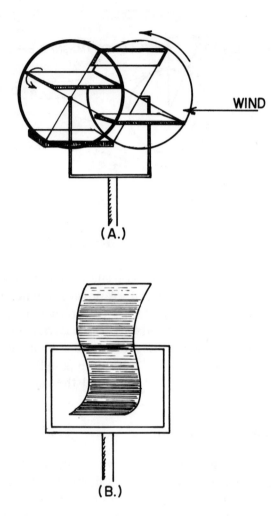

Figure 4-12. (A) Cross-wind paddle design, and (B) cross-wind Savonius design.

horizontal units when the wind direction changes. In addition, fairly complicated arrangements for collecting the output power from these systems must be engineered. This in itself is an indication of low overall efficiency.

CHAPTER 5

PERFORMANCE AND DESIGN CHARACTERISTICS

> "The wind moans, like a long wail from some despairing soul shut out in the awful storm."
> ...... W. H. Gibson—*Pastoral Days*

DEFINITIONS AND BASIC FORMULAS

By the middle of the 19th century approximately 25% of the U.S. nontransportation energy was supplied by power derived from windmills. The concept of wind energy utilization has been one of considerable interest for over a thousand years and has sparked increasing curiosity and controversy in this century. From the 1930s through the late 1950s, a series of experimental programs in a number of nations including the Soviet Union, England, Germany, Denmark, France and the United States laid the groundwork for many of the more advanced systems discussed in the previous chapter. In general, these systems are technically feasible; however, in terms of economics they are not so appealing. The capital cost per kW of a wind energy machine is for the most part high in comparison to most conventional sources of energy, with the exception of a few special applications at sites where there are high winds. One of the major parameters that will dictate the economic feasibility of investing in wind energy systems is the performance or efficiency.

For all wind energy systems, Equation 4.2 is the basic expression for evaluating the total power potential of a given wind stream. For unshrouded rotors, the fraction of that total power

70 FUNDAMENTALS OF WIND ENERGY

that can be extracted depends on the rated efficiency of the device. Mathematically, the amount of power harnessed by an unshrouded wind turbine is given by the equation

$$PP_E = \eta Q_T (\Delta \rho + \Delta K) \tag{5.1}$$

where: P_E = the power extracted,
η = the wind turbine efficiency,
$\Delta \rho$ = the change in pressure energy between the inlet and exit of the wind turbine,
ΔK = the change in kinetic energy of a unit volume of air that passes through the wind machine, and
Q_T = the volumetric flow rate of air at the turbine. The volumetric flow rate is the velocity of the wind at the turbine (u_T) times the cross-sectional area of the turbine perpendicular to the wind direction.

Consider a wind stream that is initially flowing at some steady velocity. In such a situation, a conventional wind energy system will operate under steady-state conditions, provided the wind's velocity does not undergo large fluctuations (gusts or relaxation), *i.e.*, the wind velocity is at a near-steady value. As a unit volume of air approaches the turbine, its velocity will tend to decrease monotonically. As it passes through the turbine and exits, it obtains kinetic energy from the surrounding winds. This causes the velocity to increase until it reaches its initial near-steady velocity.

The pressure energy will tend to increase as the unit volume of air approaches the rotor until a maximum pressure is reached at the turbine blade-air interface. At the same time, since the air stream velocity decreases, the kinetic energy decreases (recall from Equation 4.1 that kinetic energy is proportional to the velocity squared). When the air volume passes through the rotor, kinetic energy is imparted to the turbine blades, causing the pressure energy to decrease to a value below atmospheric pressure. Upon exiting the turbine, the air undergoes an increase in pressure until it reaches the atmospheric value. The kinetic energy continues to decrease until after passing through the turbine. Note that the sum of the changes in pressure energy and kinetic energy remain constant as the air approaches the turbine upstream.

Figure 5.1 illustrates what happens as a unit volume of air passes through a wind turbine. Streamlines of air are forced to expand around the propeller, and as they pass the turbine they tend to recede from it. The surrounding winds tend to supply kinetic energy to the disturbed streamlines, causing the effects of the turbine to diminish from the dispersal of the disturbance downstream. Note that this expansion of the streamlines causes a change in the cross-sectional area of the unit volume of air. This area is inversely proportional to the wind velocity and is the reason that the velocity tends to decrease and then increase as the air passes through the turbine.

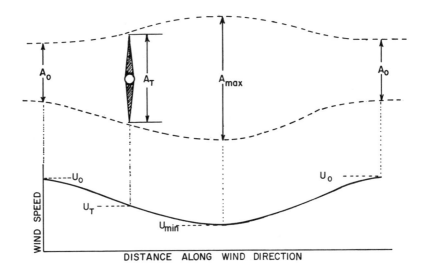

Figure 5-1. Air streamlines expand as they approach a rotor, causing a decrease in wind velocity and changes in pressure energy and kinetic energy.

At the interface between the approaching air and the propeller blades, the cross-sectional area of the wind increases from A_O to A_T (the cross-sectional area of the turbine). At this point, the velocity of the wind is reduced to roughly two-thirds of its original steady-state speed, Uo, that is:

$$U_T = 0.67\, U_o \tag{5.2}$$

where: U_o = the initial steady-state wind velocity and
U_T = the wind velocity at the turbine.

The maximum effect of the turbine on the wind streams is not felt until some distance downstream from the propeller; that distance depends on the size of the blades and the initial wind speed. At this point, the cross-sectional area of the unit volume of air increases to a maximum, and the velocity decreases to a minimum. The velocity at this point is approximately one-third of the initial velocity of the wind stream:

$$U_{min} = 0.33\, U_o \tag{5.3}$$

These observations are tied together through a parameter known as the power coefficient. The power coefficient of an unshrouded rotor is defined as the ratio of the power delivered to the turbine to the total power available in the cross-sectional area of the wind stream subtended by the propeller. Mathematically, this is expressed by the equation

$$C_P = \frac{P'}{\tfrac{1}{2}\rho A_T U_o^{\,3}} = \frac{P'}{P_T} \tag{5.4}$$

where P' is the power delivered to the turbine by the wind.

The size of a turbine is a primary factor in the available amount of power delivered to the wind machine. rotors having small total blade area produce an airstream with a high volumetric flow rate, which in turn causes a small pressure drop across the turbine. The net result, from Equation 5.4, is that the power output and power coefficient are small. For designs with large total blade area, the wind stream velocity will be small, causing a low volumetric air-flow rate and consequently a large pressure drop across the turbine. This effect will also produce a low output power rating and power coefficient. In order to achieve the maximum power output from a system, then, an optimum total blade design must be obtained that maximizes the product of the volumetric air-flow rate and the pressure drop across the wind machine.

From momentum theory, the maximum quantity of energy that can be extracted by a wind machine from a unit volume of air passing is approximately eight-ninths of the kinetic energy of the wind stream. As previously noted, the maximum reduction in wind velocity occurs at some distance downstream from the turbine and is given by Equation 5.2. For horizontal-axis machines operating at 100% efficiency, the maximum power density that can be extracted from the wind stream is given by the expression

$$\frac{P_M}{A_T} = 0.67 \, U_o \, \frac{8}{9} \frac{(\rho U_o^2)}{2} \tag{5.5a}$$

or

$$\frac{P_M}{A_T} = 0.593 \frac{\rho U_o^3}{2} \tag{5.5b}$$

where P_M is the maximum power extracted and the constant, 0.593, is known as the Betz coefficient. The Betz coefficient or limit is the maximum fraction of power that can be extracted from a wind stream of specified cross-sectional area (the fraction being 16/27).

The actual operating power coefficient varies greatly among various designs. In general, however, C_P for an ideal wind turbine is a function of the ratio of the blade-tip speed to the free-flow wind stream velocity. When this ratio approaches a value of 5 or 6, the maximum coefficient is achieved. Normally, the relationship between the power coefficient and the ratio of the blade-tip speed-to-wind-speed is displayed graphically. Such a plot is referred to as a performance curve for the wind machine. Figure 5.2 shows some typical performance curves for various wind machines.

As shown in Figure 5.2, different types of wind turbines have different optimum speeds and power coefficients. The American multiblade rotor, for example, is shown to be most efficient when the tips of the blades are moving at approximately the average wind velocity, *i.e.*, the tip-speed ratio is close to 1.0. Two-bladed rotors of the aerodynamically stable type, operating in the ideal range of 5 to 6 for the tip-speed ratio, have a high power coefficient (close to 0.48). The Dutch four-arm design is seen to be fairly inefficient, with a maximum power coefficient of about 0.18. In general, the Dutch design has high torque and low rotational speed, resulting in an inefficient blade design.

74 FUNDAMENTALS OF WIND ENERGY

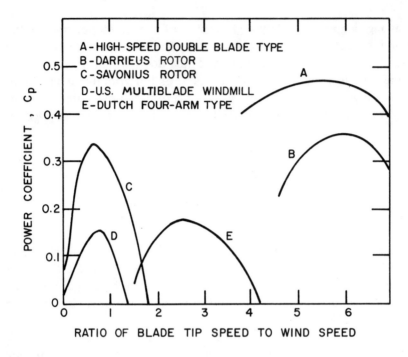

Figure 5-2. Typical performance curves for various wind turbines.

In general, the rotor blade, which is one of the most efficient among the various wind machine designs, is capable of both twist and taper action and is often equipped with a variable airfoil section. Blades of this type are generally more expensive to manufacture than are those of unchanging sections which are neither tapered nor twisted. Although there are many factors governing the economics of wind machines, which will be discussed later, the cost of blades has a significant impact on the overall economics. This by no means implies that the most cost-effective blade will be the most efficient aerodynamically. The crucial factor governing the economics is the amount of wind energy extracted per unit of capital expended. Hence, most costly blades are justified only when the value of the additional energy extracted exceeds the extra costs. In essence, then, the wind engineer is more concerned with determining the optimum system design based on the most cost-effective compromise, rather than on the most efficient design.

PERFORMANCE AND DESIGN CHARACTERISTICS

EFFICIENCY OF ELECTRICAL GENERATION SYSTEMS

When wind speeds fall below the turbine rated speed, the rotor must be capable of varying with the wind velocity so that the maximum possible amount of power can be extracted. As pointed out earlier, this does not match the optimum operating conditions of either the induction or synchronous a.c. electrical generators. The concept of interfacing the rotor with its electrical output has been developed for this reason. As noted earlier, one approach is to make arrangements such that the rotor varies optimally with the wind velocity. In such a case a variable-speed constant-frequency generator could be employed in obtaining the steady frequency power required for the electrical facilities. This can be accomplished by differential systems which are mechanical or electrical frequency changing devices, or by nondifferential systems which are static frequency changers or rotary units.

Before covering the highlights of wind machine performance with regard to electricity generation, it is important to understand the basics involved in the mechanical energy generated from the wind turbine. Consider a horizontal-axis rotor blade having a fixed pitch angle. The torque developed by the rotor depends on both the rotational speed of the shaft and the wind velocity. If during a relatively steady wind speed the rotational blade speed is too low, the blade will stall. This in turn causes the output torque of the wind turbine to decrease. In order to derive the maximum power output as the wind speed varies, either the pitch angle of the blade or the blade's rotational velocity must vary. Hence, many of today's aerogenerators have included in their designs variable-pitch blades. These variable-pitch blades are adjusted to operate at as near a constant rotational speed as possible, for wind speeds that go above a minimum (*i.e.*, a cut-in value) and for varying output loads. A family of curves can be generated showing how the blade's torque varies with the shaft's rotational speed for different wind velocities. Curves showing when the blade stalls and when maximum power can be extracted can be superimposed over this plot to provide design information on the output performance from prototypes. Figure 5.3 shows such a plot.

We have already seen in Chapter 4 how the power output or efficiency of a wind generator can be markedly reduced by

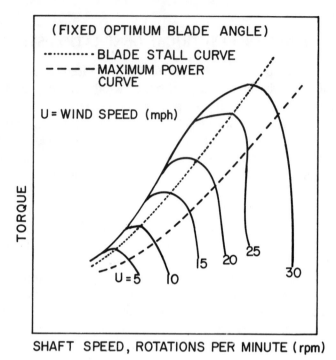

Figure 5-3. Generalized speed-torque plot.

interfacing the wind machine with an electrical generator. Another factor which affects efficiency, although its effect is small, are friction losses within the operational range of the aerogenerator and the a.c. generator. It is important to note that designs should be as close as possible to their maximum tip-speed ratio. Hence, there must be a trade-off between designing for a match with the a.c. electrical generator and the aerodynamic parameters involved. Figure 5-4 illustrates that no actual wind machine ever achieves its theoretical efficiency, even when operating at its optimum tip-speed ratio. The actual output curve may be constrained to operate at a constant shaft rotational speed because of the interfacing requirements for the synchronous a.c. generator, but the efficiency curve is shown to decrease sharply as the tip-speed ratio departs from its optimum value.

Of course, the primary factors affecting aerogenerator efficiency are the wind variability and average energy available at a specified site. Power density duration curves aid the wind

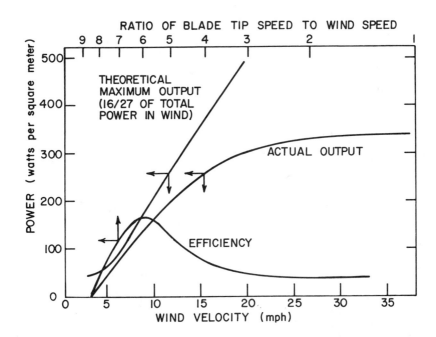

Figure 5-4. Illustrates how an ideal horizontal-axis wind generator's efficiency is affected by interfacing requirements with an a.c. synchronous generator and how this in turn affects the tip-speed ratio.

engineer in assessing the amount of time that a given site will provide a specified amount of energy to a wind turbine. That is, such curves indicate the operation or run time of a wind generator. Figure 5-5(A) shows a typical power density duration curve for a specified site. The total amount of energy per year available in the wind per unit swept area of the wind turbine is the area under the power density curve, in units of kilowatt hours per square meter per year ($kWh/m^2/yr$).

The actual power density of an aerogenerator coupled with a synchronous a.c. generator driven by a gear train with a constant or fixed ratio would be considerably less than the total available or theoretical power density illustrated by the area under Figure 5-5(A). The actual power density would depend on the cut-in wind speed, the rated wind speed of the wind machine and the cut-off wind speed (a cut-off speed is used to shut down the wind machine when winds are sufficiently high to cause damage

78 FUNDAMENTALS OF WIND ENERGY

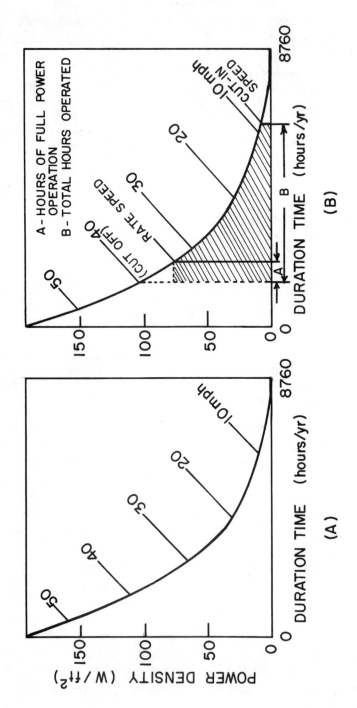

Figure 5-5. (A). Power density duration curve for a potential wind site. (B). The actual annual power density output for a wind energy system.

to the electrical generator and/or blades). Figure 5-5(B) illustrates that the actual power density extracted by a system is considerably less than the total available. As an example, if a system is designed for a cut-in wind speed of 20 mph and a cut-off wind speed of 40 mph, and if the rated wind speed of the wind turbine is 33 mph, then the shaded area under the power density duration curve in Figure 5-5(B) represents the actual output power density per year of the system.

There are a number of parameters that can be altered to improve efficiency. Two parameters which have a strong effect on the power density are the swept area of the rotor and the wind velocity. Although wind velocity is more difficult to manipulate, it can be somewhat specified to improve design by the proper selection of a wind energy site, which will be discussed in Chapter 6. As has already been shown, power (or power density, P_T/A, where A is the swept area of the rotor) will increase as wind speed increases. From Equation 4.2 it can be seen that for a 15-mph wind the maximum power that could theoretically be extracted from the wind is approximately 25.7 W/ft^2. However, if our system is located at a site having an average wind velocity of 30 mph, then the power density is increased to 126 W/ft^2 of blade area, or roughly 8 times more power is associated with the new site. More available energy means a wider operating range for the WECS.

Increasing the diameter of the rotor can also increase efficiency or available power. For the example above, if the diameter of the swept area of the rotor is increased from 10 to 20 feet, this produces an increase in swept area of 78.5 ft^2 to 314 ft^2 ($A = \frac{1}{4}\pi D^2$). At 15-mph wind speeds, then, the available power is increased from 1.23 kW to 4.93 kW over the two swept areas, or doubling the swept area diameter has the effect of increasing the total power in the wind stream by a factor of four.

It should be noted, however, that the wind engineer's job is not as simple as the correlations and graphs presented in this chapter make it appear. The specific size and location of a wind machine must be based on an optimum design. That is, these parameters must be selected on the basis of minimizing costs to the consumer and of the energy produced. Numerous complicated factors such as transmission and distribution costs of proposed systems, in addition to physical limitations on the rotor blade size and tower structure must be accounted for.

LARGE-SCALE DESIGN CONSIDERATIONS AND ECONOMIC FACTORS

For electrical energy generation—and in general for most wind energy systems—the major design components include the rotors, the type of transmission requirements and arrangements for transmitting torque from the rotor to the generator, the generator capable of producing electricity at suitable line voltages and phases, the control systems for orienting the rotor in the wind direction, control systems for monitoring the generation of electricity and protecting the wind energy machine components from damage under abnormal conditions such as high winds, and the tower support for the system components.

It has already been shown that the power output varies with both the wind swept area of the rotors and the cube of the wind velocity. Because of this cubic relationship and the fact that the wind speed is unsteady, the actual power output of a WECS over an extended period of time may indeed be a great deal larger than the estimated power based on average wind speed data over that time period. In addition, the height at which rotors are located will also determine the output characteristics. Hence, the tower height may be a significant fraction of the total cost of such a system.

Requirements for the size and power output of a wind energy system may be dictated by limitations on the strength of materials. This in turn can affect the size requirements of blades, bearings, tower heights, etc., which enter into the economic picture. Consequently, if a large capacity wind energy system is required, it may have to be composed of several interconnected, dispersed wind turbines that are relatively small.

Applications of wind energy systems are categorized in terms of the type and amount of power output needed. The requirements of the consumer are foremost, as an uninterruptible power supply is most often demanded. The energy supply can be interrupted in a number of specialized applications, such as grinding operations, irrigation or watering livestock, or in applications involving the production of electrolytic hydrogen for fuel storage. To make WEC systems attractive, particularly to utilities where they might find greatest use, prospective sites will require back-up power systems to maintain a continuous supply of electricity during periods of inadequate windspeed. Such back-up units may include gas turbines or diesel engines. For supplying base loads that are not equipped for interconnection to conventional systems, on-site energy storage will have to be used. The most common energy storage systems include:

(a) batteries or stored electrolytic hydrogen,
(b) mechanical storage such as compressed air and pumped storage,
(c) magnetic storage such as inductive coils having no resistance, and
(d) thermal storage such as sensible-heat storage by capacity and/or physical changes that involve heat, fusion or evaporation.

Seasonal changes in wind variability may make such storage system arrangements economically appealing. The additional cost associated with energy storage can be minimized by utilizing existing capabilities such as depleted natural gas wells or other underground storage structures for housing hydrogen and other compressed gas fuels.[16]

The choice between using auxiliary conventional power generating systems and various storage system approaches must be based on a thorough economic analysis. In addition, the wind engineer must analyze the specific power requirements and available geological conditions at the chosen site. One final aspect that must be carefully examined is the possible environmental impact that one system may have over another. This may also enter into the economic analysis if environmental control systems must be incorporated with the auxiliary power supplies or storage systems used.

The ultimate objective in designing a wind energy system for a particular power application is to minimize the cost of a system over its intended life while at the same time to maintain the cost of energy production in a competitive or lower range than conventional approaches. To achieve this goal, the capital cost or investment in a particular system must be kept down to a minimum. This involves careful economic study of the energy storage system required and proper design for blades, generator, gears, bearings, tower, etc., to ensure low maintenance and operation costs during the system's life cycle. Included in the economic studies are careful evaluation of the energy payback time. The payback time is the period required for the system to generate sufficient quantities of energy from the wind to match the energy expenditure in manufacturing, operating and maintaining the wind energy system during this allotted time period. Another criterion for evaluating the economic feasibility and efficiency of a WECS is the energy gain which is defined by the following formula:

$$G = \frac{P_T}{P_{ex}} \qquad (5.6)$$

where G is the energy gain (dimensionless) which is the ratio of the forcasted total amount of power production of the wind energy system for its predicted lifetime (P_T) to the predicted amount of power expended in manufacturing, operating and maintaining such a system over its life-cycle (P_{ex}). Obviously, such economic and performance predictions for large-scale systems must be based on detailed experimental studies on smaller prototypes. There is, however, a large degree of uncertainty in scaling large systems up from smaller experimental units.

The methodology for minimizing energy production costs for conventional wind energy systems has been developed by Golding.[17] The interested reader is referred to this work for a more detailed dicsussion. As Golding points out, although the capital cost of wind generators increases with the rated output, size and complexity of the system, the capital cost per kilowatt actual decreases. Costs associated with the interconnection and transmission arrangements may also decrease as the size of a unit increases (this assumes that the ultimate design is within the limitations of the materials of construction, such that fewer units may be required for a given application). Another point of interest is that the capital cost tends to decrease as the rated wind speed of the system increases. When units are designed for higher rated wind speeds, the size and weight requirements for the rotor, blades, gears, etc., becomes smaller for a specified output capacity. Golding's analysis has shown that if large-capacity units in the MW range are mass produced, cost estimates are roughly half of the medium-size units in the 100-kW range.

One deterrent, however, is that the load factor of a wind turbine tends to increase as the average-to-rated wind velocity increases. During the operation of a wind machine, various extraneous loads are imposed on the system, which can cause excessive wear and affect the life-cycle of the unit dramatically. These loading effects are caused by a variety of conditions such as:

(a) the force of gravity on the rotor blades when they are in a horizontal position,
(b) tower shadowing, whereby a portion of the blade is concealed by the tower structure resulting in an uneven distribution of impact force on the unexposed blade swept area,
(c) wind gusts which can cause bending or distortion of rotor blades, and
(d) sudden shifts in wind direction which can cause an uneven load distribution on the blades.

These conditions can cause cyclic motions and/or variations in the blades on conventional wind turbines, or may generate severe vibrations on the support tower, the bearings, or elsewhere. Figure 5-6 illustrates the causes of some of these loading effects. Additional design considerations and selection of construction materials can have a significant effect on the economic feasibility. The wind engineer is therefore faced with further optimization problems between equipment costs and load limitations in achieving a low-energy-cost wind energy system.

Outlining proper economic methodology, forecasting techniques and the numerous cost factors necessary in evaluating a WECS is worthy of a book unto itself. Capital costs are by no means the only aspect that must be carefully examined. Changes in interest rates or in the cost of money can have a significant effect on the total economic view. Just as important as the actual cost of manufacturing a wind machine is the ability to predict its manufacturing, maintenance and operating costs. To make

84 FUNDAMENTALS OF WIND ENERGY

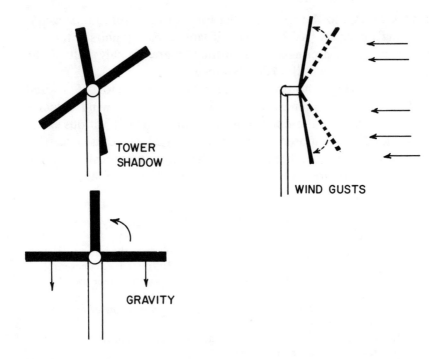

Figure 5-6. Extraneous loading conditions can cause excessive wear to wind machines, resulting in further design considerations to ensure an acceptable operating life.

systems appealing, there must be considerable reliability in economic forecasting or else the ultimate consumer of a wind machine will not want to invest in such a program.

CHAPTER 6

WIND SITE SELECTION FACTORS

"Nature tells man to consult reason, and to take it for his guide."
.......Paul H. T. D'Holbach
The System of Nature, 1770

WIND DATA SOURCES

In general, there are two possible siting locations for wind energy systems: on land and offshore. Land-use facilities are usually considered more permanent sites. The ultimate selection of a wind energy site is a lengthy procedure that is based more on intuition and good judgment than on a methodical, exact set of guidelines. Extensive analysis of meteorological data and wind characteristics is necessary. Unfortunately, currently available maps that show patterns of average wind power over large regions provide only crude estimates of this power. In Chapter 4, it was shown that wind power has a cubic dependence on wind speed. Hence, detailed information on climatological distribution of speeds is necessary for calculating averages of wind power.

The National Climatic Center in Asheville, North Carolina, is reponsible for collecting and organizing wind speed distribution data in the continental United States. Approximately 700 meteorological stations provide annual average sea level wind contours to the Center. Similar information is prepared for seasonal variations. Figure 6-1 shows rough annual average wind contours across the country. Comprehensive observations of various wind

86 FUNDAMENTALS OF WIND ENERGY

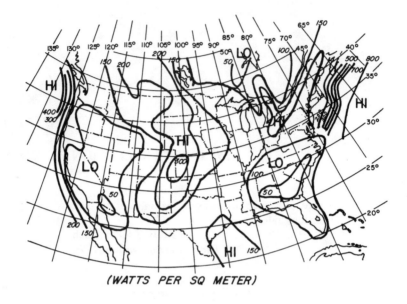

(WATTS PER SQ METER)

Figure 6-1. Annual average wind contours in the United States.

characteristics from over 2000 locations are fed through the various meteorological stations to the Center. Observations of temperature, pressure, wind direction, wind speed and cloud coverage are usually recorded once each hour, primarily during adverse or abnormal conditions. At approximately 700 of these locations, measurements are made 24 hours a day while at the remainder, information is recorded every 3 to 6 hours during daylight. The latter information is primarily useful for local air traffic requirements. Hourly wind speed values are evaluated by averaging speeds observed for a one-minute inverval on the hour. Gusts are defined as velocities which exceed the lowest recorded speed during the time interval by a minimum of 10 nautical miles per hour. Data reported to the Center is usually in manuscript form.

All meteorological data covering land and marine areas that are recorded by the National Weather Service, the military services and the Federal Aviation Administration (dating as far back as the latter part of the 19th century) are filed at the Center. The bulk of the data recorded following the end of World War II is digitized and is available on punched cards and/or magnetic tapes. All hourly observations, up to 1964, except for gusting information are available in digitized form. Marine data recorded by the U.S. Navy and various domestic merchant marine vessels are also in the Center's archives. Summaries of annual and monthly evaluations are provided as part of the Center's function. These summaries indicate the occurence of wind in various speed classes along with the wind direction information by frequency or percentage frequency.

Marine data collected by ships are also available in summary forms. This information is usually characterized by geographical areas on latitude-longitude grids. Summaries of immediate coastline wind climatology can also be obtained, although the volume and reliability of these data are limited as most vessels discontinue their observations within 50 miles of land.

Wind gust data can often be obtained from the original manuscript record of the testing station. Many stations are equipped with continuous wind velocity recorders which will supply a printout from which peak gust information can be extracted.

Many testing stations conduct routine observations of wind characteristics aloft. These studies are made by tracking a rising gas-filled balloon with tracking antenna. Most of these studies are conducted twice daily and data are usually recorded at 150 and 300 m above the ground. A few specially designed instrumental towers conduct wind profile studies within the lower 150-150-m range of the atmosphere.

DATA QUALITY

Although a massive volume of historical and statistical information exists on wind characteristics, its usefulness in wind energy technology is limited. There are number of aspects about wind climatology that are poorly understood because of lack of data. To begin with, the bulk of the data that has been accumulated since the 1940s was observed from airport sites and residential and city locations. It is highly questionable whether the criteria employed in airport siting can be applied to wind energy turbine site selection. In rough terrain such as hilly regions, observation stations are normally located in flat areas or in valleys where the wind strength is considerably less. As will be shown shortly, local topography has a significant effect on wind direction and speed in a valley. Little information is available to the wind engineer on wind conditions at hilltops.

Another problem making it difficult to assess the Center's information banks is the lack of standardization in recording instruments. The National Weather Service (NWS) has always employed the vertical-axis rotating-cup anemometer for wind velocity measurements; however, many independent research groups and portions of the military services employ other instrumentation such as sonic, hot-wire and/or pressure-plate anemometers. Early instruments employed by the NWS for monitoring wind speed were classified as totalizing. The totalizing type recorders monitored the passage of a number of miles of wind past them over a specified time interval, from which speeds were directly recorded in mph. Although direct-reading instantaneous instrumentation is available today, many of the NWS measuring stations still employ totalizing units. Such systems are not capable of recording short-term wind fluctuations and gusts; hence, a great deal of information is not available at many potential wind energy locations. Furthermore, because of the variety of techniques available for monitoring wind characteristics, the engineer is faced with an even more difficult task of interpretation. It should be noted that the NCC does not archive information obtained from instruments other than the so-called standard rotating-cup anemometer, and the wind engineer has no basis for comparison of various instrumentation precision and accuracy in measurements. Coupled with this problem is one caused by the variability of anemometer measurement heights. Anemometer heights vary from 6 to 60m above the ground at various measuring stations, and very little of information has been sorted by the NCC.

Little or no information exists concerning height effects on wind velocity and direction. Free air traveling above the boundary layer flows at nearly twice the recorded standard-height anemometer velocity of air flowing over flat ground or water surfaces. A rough estimate is that wind *velocity* increases with the one-seventh power of height and wind *power* increases to the three-sevenths power of height. This is a fairly good rule of thumb, although more accurate criteria and correlations have been developed for the wind-speed-height relationship.[18,19] The major problem, however, is that many of these empirical correlations are based on measurements taken over fairly flat terrain. Scarcely any data have been recorded over rougher topography, such as over mountain ridges or even high-rise buildings.

The wind engineer is lacking the most information, probably in the area of the turbulence structure of the lower atmosphere. Although measurements have been made, turbulence observations are generally not routinely monitored at reporting stations, and no substantial data bank exists in the NCC's files. Turbulence energy spectra are badly needed as a basis for design of wind energy conversion turbines. Of primary importance is the longitudinal wind spectrum. The complex dynamics of the atmospheric boundary layer are not understood. Few numerical and physical models have been developed that attempt to describe heat and momentum flows in the atmosphere. Observations that have been made indicate that the characteristics of wind spectra can vary greatly depending on the terrain. Hogstrom[7] has proposed a wind spectrum scale on the basis of roughness elements. He claims that the shape and arrangement of roughness elements are a primary factor in the turbulence structure. As an example, Hogstrom points out that the turbulence structure over a homogeneous terrain appears different from that terrain characterized by alternating open fields and patches of forest, even though the overall roughness of both terrains might be the same.

There is a real need for better estimates of median wind speeds and of wind duration profiles (*i.e.,* cumulative frequency distributions of wind speeds). These improved estimates can only be made from more detailed observations and better wind profile correlations. Studies have shown that there is some optimum height to be evaluated in wind turbine construction. That is, the designer must determine an optimum height at which contributions from the mean wind speed and the shape of the wind duration profile provide a favorable diurnal variation. The optimum height will, of course, depend on the climatological and physical properties of the potential site.

Some information is available on gustiness characteristics; however, there is a definite need for more research in this area. More detailed data on peak wind speeds and duration times are needed for adequate assessment of potential wind sites. This information is also required in evaluating the effects of coupling geographically separated groups of aerogenerators that would be used in large-scale energy production. Some of this information can be estimated from statistical data on atmosphere stability and mixing height effects. In addition, ventilation indexes for air pollution dispersion estimates are applicable. It should be pointed out that long-term historical data (a minimum of 5 years recorded observations for a potential site is necessary) should be used in site evaluation. There is a great deal of uncertainty and danger in basing a site selection on short-term observations without detailed comparisons being made to individual weather periods relative to average climatic conditions.

CRITERIA FOR SELECTION

The wind engineer must ask three questions before selecting a wind energy conversion site:

(1) Where does one look for potential sites?
(2) What specific data are necessary for evaluating a particular site?
(3) If measurements are to be made because of the lack of existing data, over what period of time should they be taken?

The importance of selecting the proper site for a WECS cannot be overemphasized. The choice of a site based in incorrect decisions or insufficient wind characteristics data can be the single most important factor leading to the failure of a system.

The basis for selecting an optimal site stems from careful evaluation of wind characteristics at potential sites from long-term measurements at nearby measuring stations and limited short-term observations at the potential site. As noted earlier, wind velocity is the wind property for energy conversion. Records of long-term wind variations from measuring stations near a potential site can supply sufficient information to construct a fairly accurate description of the local wind velocity. This source of data should be used, however, only to provide background information and to confirm measurements taken at the actual site considered, as there may be topographical differences producing different wind characteristics. This type of information is useful primarily in that the same convective systems will determine the wind direction and strength in both places. Topographical features, short-term data collection at the potential site and long-term data from sites in the same proximity can often provide reliable estimates of wind characteristics at the considered wind energy location.

There are a variety of approaches to evaluating wind characteristics and much work is being done in the area of mathematical modeling to describe and predict daily wind averages. There has been some success in describing hourly variations in wind velocity with zero-mean Gaussian distribution functions.[8] This type of information is particularly useful in predicting consistency and variability of power output from an aerogenerator. Similar success has been achieved with modeling wind direction using beta distribution functions, and other methods.

Although field studies provide the most reliable and detailed information, they are also time-consuming and costly to conduct. Ideally, a wind survey should be conducted in a minimum amount of time without being a significant portion of the project cost, and it should have reasonable prediction reliability. There are a number of crude estimates, based on field measurement studies, that enable the wind engineer to obtain qualitative guidelines for selecting sites.[20-23] These studies are primarily site-specific and have little general applicability; however, general trends can be predicted. Again, the major problem with using this information is the lack of detailed observations of wind properties over rough terrain.

Trees, buildings, hills and even wind machines affect the wind characteristics or, more specifically, the wind shear. Figure 6-2 illustrates how wind shear is affected by different roughness elements on the earth's surface. In general, relatively rough terrain results in low winds under 60 m. Hence, taller support towers for wind energy converters would be required.

The influence of topographical conditions on wind power can be classified in four broad areas:

(1) variations in wind velocity over flat or uniform terrain,
(2) flow of wind over slight to moderate roughness elements,

92 FUNDAMENTALS OF WIND ENERGY

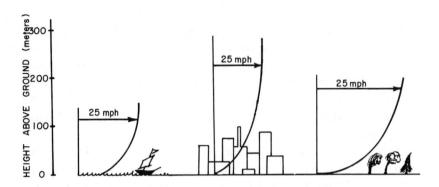

Figure 6-2. Ground roughness elements can distort the wind velocity profile.

(3) flow over mountains and ridges, and
(4) local wind currents and circulation.

Class 1 has been studied extensively, and a large volume of data exists for this category. It should be noted, however, that sudden changes in roughness or temperature even over relatively flat areas can generate significant alteration in the wind's climatography.

Slight or moderate roughness elements cause winds to overshoot, *i.e.*, wind velocity tends to increase. An example of this condition is flow over a small hill. When wind streams pass over a hill they are compressed and the flow is accelerated on the leeward side, as shown in Figure 6-3. This effect can be beneficial to wind turbine operation and is a condition the wind engineer would look for in examining sites. Frenkiel,[21] Golding,[16] Plate and Lin,[24] and others[25,26] have recorded wind speeds and other measurements over hilltops. Unfortunately, there are no prediction schemes available at present for determining even crude estimates of wind velocity over hill summits. Stratified shear flow has often been observed over hilltop situations, which can have a very significant effect on the flow. In addition, the intensity of inversion layers can influence wind speed and direction. In general, if inversions frequently occur, the hillside experiences greater annual wind speeds rather than the hilltop.

High mountain ridges greatly affect wind climatology. Winds flowing over high mountain ridges are often highly unstable and subject to gustiness. Furthermore, stratified flow over mountains

WIND SITE SELECTION FACTORS   93

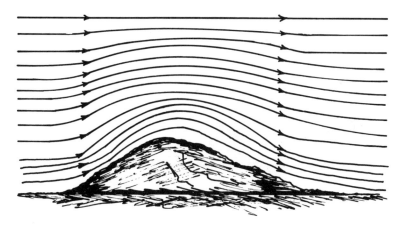

Figure 6-3. Winds passing over a hill are accelerated.

can give rise to abnormalities called lee waves or helm winds. Besides high winds, mountain ridges can produce areas of underspeed or reverse flow. Figure 6-4 shows a condition that often arises on a hill or mountain ridge with abrupt sides. As shown, the wind velocity can be greatly reduced and part of the flow may reverse direction. Obviously, such a site would make a poor selection.

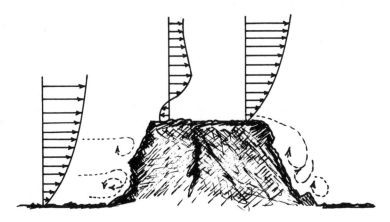

Figure 6-4. Ridges with abrupt sides can cause reverse flow and slow down velocities. Such a condition is unsuitable for a wind energy conversion site.

Mountains or ridges with sharp peaks can, however, produce favorable conditions for a WECS. Figure 6-5 shows such a case. Here, the velocity profile is redistributed in such a manner as to

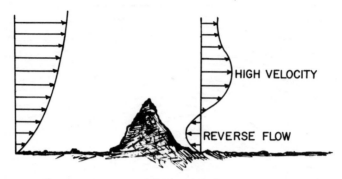

Figure 6-5. Sharp peaks can produce favorable wind energy sites.

create a region of high speed at some height above the ground. It is also possible that reverse flow may develop. If such a site is selected it is essential that the aerogenerator be positioned at exactly the proper height to match the velocity profile.

Although mountain sites are generally associated with variable winds, they may be the best potential sites. Winds are generally stronger with greater velocities, and much more power is available at these locations. If costs can be limited, further refinement of WEC systems may produce suitable designs for such sites. Another possibility is to alter the natural surroundings by creating man-made ridges or removing roughness elements. Figure 6-6 shows an example of how this might be done for a particular situation. If a narrow channel is excavated between two mountain ridges, the wind will become more uniform and tend to speed up the passageway (wind flow is into the diagram). Hence, stronger, more uniform wind velocities may develop, making a potentially good wind energy site.

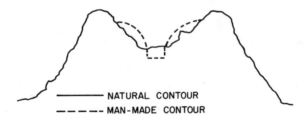

Figure 6-6. Man-made changes to the terrain can produce artificial wind energy sites.

The final class of conditions that the wind engineer might encounter in his search are local wind circulations. Very little information is available on these, although many qualitative generalizations have been made from simulated tests in wind tunnel facilities. An example of local circulations would be those winds experienced along shorelines. Steady local breezes can greatly influence the atmospheric motion of the wind, quite often favorably. In general, they create appealing sites. Natural passes or gaps, as shown in Figure 6-6, can create enhanced situations, particularly if they are in the direction of prevailing winds. The primary driving forces behind local circulations are surface roughness, temperature gradients or pressure (or density) gradients in the atmosphere.

GENERAL APPROACH TO SITE SELECTION

The various factors, situations and types of meteorological data discussed in the preceding sections must undergo careful evaluation during the selection process. In addition to these many parameter, one final criterion must be met before a final choice is made; that is, the ability of the potential site to provide economic amounts of wind energy. The assessment of this point must be based on climatological evaluation of expected performance over the anticipated life of the wind energy system. Hence, the site for a WECS plays a major role in the economic outlook of such an investment.

In the United States, sites that have sufficiently strong and relatively steady winds are few. Many of the most promising locations are removed from energy consumers or existing utility stations. Hence, a major problem in this country is to locate high wind regions which provide assurance of fairly steady winds, which would lend reliability to predicted performance characteristics.

The general approach to evaluating a site includes:

(1) making a survey of all historical meteorological data for a particular region of interest,
(2) preparing and examining contour maps of the terrain,

(3) based on statistical analysis of average wind speeds, direction, environmental conditions, etc., narrowing the choice down to several potential sites,
(4) monitoring meteorological conditions at the most promising sites for a minimum of one year.

An exact methodology for narrowing the choice of locations does not exist at present. The wind engineer must rely heavily on meteorological background information, on general guidelines and rules of thumb, and on a limited number of field studies from site visitations. A field meteorologist with local experience is a worthwhile consultant. (Qualitative guidelines and a theoretical background to wind energy sitings are discussed in detail in a World Meteorological Organization publication.[18]) Figure 6-7 outlines the general procedure that can be used in narrowing down the selection.

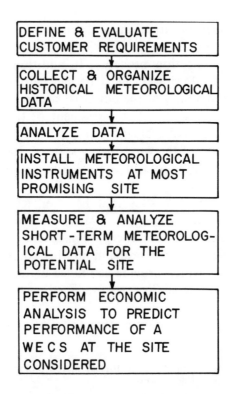

Figure 6-7. General approach to selecting a wind energy conversion site.

Mathematical models would greatly simplify the selection procedure. Mathematical model-based meteorological prediction schemes have been attempted fairly recently.[26,27] Meteorological simulation techniques have been applied successfully in modeling synoptic-scale weather. These models are capable of predicting conditions over 200-km between synoptic weather stations. In predicting conditions for a specific site, much smaller domains (on the order of 1 to 2 km) are considered. These techniques are unfortunately still in their infancy.

It is feasible that such numerical techniques can soon be developed. Accurate forecasting of windmill sites is of prime importance to the economic success of WEC. Consequently, any ultimate model to the developed must be tailored to account for energy consumers in different regions with different needs. Clearly, extensive field studies on particular sites is an inefficient approach to selecting an installation home. This in essence involves solving the same problem over and over again for each new potential site. Selection methodology must be better defined and standardized. This means that more carefully planned experimental field studies must be made to correlated meteorological conditions and to provide a more physical understanding of the turbulent nature of our atmosphere. Only with this information and through the use of reliable mathematical models can we ensure the success of a profitable WEC program.

CHAPTER 7

ENERGY STORAGE SYSTEMS AND ENVIRONMENTAL CONSIDERATIONS

"He that troubleth his own house shall inherit the wind." .... *Proverbs* 11:29

OVERVIEW OF WECS

It is worthwhile to review briefly some of the system characteristics that make WEC an appealing source of energy production, as well as some of the disadvantages that make it unfeasible at present. To begin with, the wind is a clean, replenishable source of energy, and although it is intermittent and relatively dilute in comparison to nuclear and fossil fuels, it constitutes a vast and practically untapped natural energy resource. There are a number of technically feasible applications of wind energy; of particular interest is electric energy generation along with various energy storage arrangement capabilities. Economic studies indicate that base-load WEC systems utilizing compressed air or hydrogen storage facilities in wind speed ranges of 15 to 18 mph are expected to be competitive by 1990 with conventional fossil-fuel systems using fuel oil costing the equivalent of $10 to $11 per barrel (based on 1975 dollars). The primary geographical regions in the U.S. where such systems appear promising include New England, the Atlantic seaboard above Cape Hatteras, upper New York State, the Texas panhandle, the high Great Plains east of the Rocky Mountains, the Sierras, and coastal regions of the

100 FUNDAMENTALS OF WIND ENERGY

Pacific Northwest, Alaska, Hawaii, as well as Puerto Rico, the Aleutian Islands and the Virgin Islands. By way of review, the WECS fuel cycle stages are: wind energy extraction, conversion to another form of energy (*i.e.*, electrical), energy storage and energy consumption. Figure 7-1 illustrates the cycle.

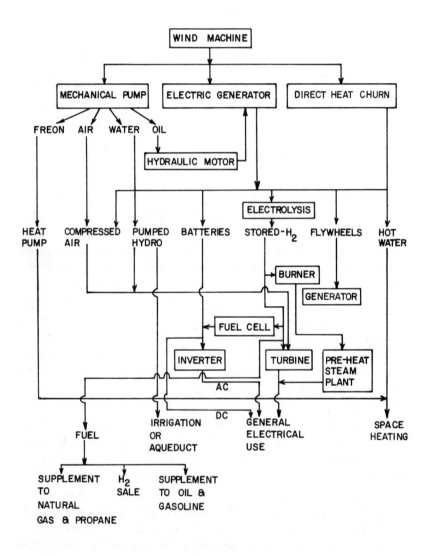

Figure 7-1. WECS fuel cycle.

ENERGY STORAGE SYSTEMS 101

It is anticipated that by the 1990s WEC systems will be capable of producing the following energy products:

(1) mechanical energy,
(2) electrical energy,
(3) heating and cooling buildings and homes,
(4) process heat,
(5) fuel gases such as hydrogen,
(6) oil and gasoline supplements such as alcohol, ammonia, etc.

Earlier, the basic taxonomy of wind energy systems was discussed. Recall that there are many different types of horizontal-axis WECS turbines that have already been developed for operation in ambient wind streams. These systems are designed to rotate at relatively low speeds, in the range of 20 to 100 rpm. In general, these systems create downwind disturbances within 5 to 20 turbine-diameters from the turbine. These disturbances include small pressure drop, various turbulence and some vortexing at the blade tips. A modern prototype of a horizontal-axis turbine is shown in Figure 7-2.

Figure 7-2. Modern horizontal-axis WECS turbine.

The various vertical-axis WECS turbines are also designed to operate in the speed range of 20 to 100 rpm and are thought to create wind stream disturbances at about the same downwind distance from the turbine.

One particularly interesting concept that has not been discussed so far is the vortex generator. The vortex generator has been proposed in the form of specially designed buildings, towers, baffles and a variety of other structures that spin the wind stream, thus creating strong vortexing at the wind turbine. The effect of these structures on the wind stream is an increase in the speed of the air flow through the nearby wind converter's power output. The types of vortex generators include unconfined and confined vortex, and horizontal- and vertical-axis rotors.

Unconfined vortex systems employ wing-like structures to deflect the wind and generate an unconfined vortex around the turbine. Estimates are that such systems can produce up to six times more power output than many of the more conventional designs discussed earlier, with the same rotor diameter.

Confined vortex systems, currently being developed by Grumman Aerospace Corporation, are devices in which the pressure drop across a ducted turbine and the wind speed through it are further enhanced by the use of additional ambient wind to generate a confined tornado-like vortex in a tower positioned near the exit of the duct. A variety of structures has been suggested to create confined vortices by capturing the wind over a wide area and spinning it as it enters an enclosure that is connected to the exit of the duct in which the turbine is housed. A circular tower could serve as the structure. The diameter of the tower might be three to four times the turbine diameter, and the tower height roughly three to four times the tower diameter. It should be noted that such systems would probably create large wind stream disturbances such as large pressure drops in the core of the vortex and rather high wind speeds in the core. Figure 7-3 shows an artist's conception of one type of vortex generator design.

As has been suggested earlier, there are many different potential WECS applications. Essential to many of these is an efficient economical energy storage arrangement. Many of the applications discussed are small scale, and there does not seem to be a need to develop technology for the economical storage of high-quality

ENERGY STORAGE SYSTEMS 103

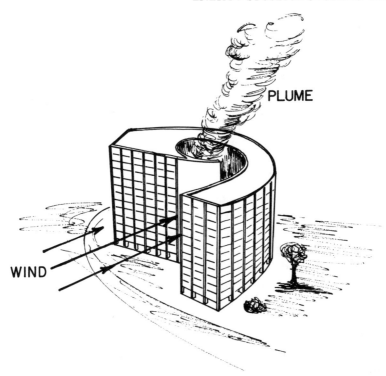

Figure 7-3. Artist's conception of WECS vortex generator in the form of an office building.

energy in large quantities. The exception to this is large-scale electric energy generation. The purpose of developing technology for storing energy is to make the combined WEC and energy storage systems an economical unit that functions as an independent base-load power supply for a wide range of conditions as dictated by the consumer.

ENERGY STORAGE SYSTEMS

There are various types of energy storage systems. Among those considered most adaptable to WEC systems are batteries, flywheels, pumped-hydro, compressed air, hydrogen storage, thermal and electromagnetic.

Electrochemical Energy Storage

Storage batteries are devices that deliver electric power by electrochemical reactions that take place between two electrodes immersed in an active solution. The solution is referred to as the electrolyte. Batteries can accept and store electric energy in the form of chemical energy by reversing the chemical reactions. The chemical makeup of a single battery cell determines its voltage, and desired voltages are achieved by connecting cells in series.

The specific energy storage of this type of battery is expressed as watt-hours per pound (W-hr/lb) of total weight. Its specific power is the rate at which it can deliver energy. It is of interest to note that the battery's specific energy goes down as the specific power increases, due to the battery's internal resistance. A storage battery's life is measured by the number of recharges or cycles it can undergo.

There are many different types of batteries that are commercially available. The reader is probably most familiar with the automobile starting battery. This is a lead-acid system that has a life normally around 200 cycles and an energy delivery capacity of about 15 W-hr/lb at low output. Lead-cobalt batteries are another example; here, cobalt salt is added to the positive plate which increases the energy delivery from 15 to 18 W-hr/lb.

Edison cells are one of the earliest known modern batteries that consist of nickle-iron. These are generally associated with longevity (around 3000 cycles); however, their energy storage and power output are below lead-acid systems.

Nickel cadmium batteries have a life of about 2000 cycles with a limiting energy density of 15 to 22 W-hr/lb. Power densities can range anywhere from 160 to 350 W/lb. These batteries are, however, fairly expensive because of the limited supply of cadmium.

There are a number of relatively new battery systems that might be adaptable to WEC arrangements. Zinc-air batteries have recently been developed into rechargeable units. This is done by circulating the electrolyte with a pump, resulting in a high recharging rate. Such systems have high specific energy delivery in the range of 60 to 80 W-hr/lb.

Organic electrolyte batteries are another possibility. They employ lithium or sodium as one electrode, in the absence of water, with certain metals. Care must be taken in selecting the proper metal for the cathode (cathode is the positively charged electrode; anode is the negatively charged electrode), as these salts will react exothermically with certain metals. Copper chloride, nickel chloride and nickel fluoride have been suggested in use with propylene carbonate as the electrolyte base.

Fused salt electrolytic batteries use a molten lithium chloride electrolyte. The battery systems described above are capable of operation at near ambient temperatures; however, lithium-chloride batteries operate at 600 to 700°C. Molten lithium-chloride and gaseous chlorine must be encased for long time periods in a suitable housing that can provide adequate electrical insulation as well as being air tight.

Sodium-sulfur batteries are another prime candidate for heavy-duty energy storage for a WECS. These units also operate at elevated temperatures, in the range of 250 to 350°C. Liquid sodium and sulfur combine to produce Na_2S_3 in a solid electrolyte of beta-alumina. Both the power and energy density of these systems are high; furthermore, their permitted changing rate is comparable to the maximum discharge rate. The cost estimate of these systems is between $20 and $40 per kWh.

There are several other systems that are currently being developed. In general, the state-of-the-art with regard to batteries is reasonably well developed. Main research interests lie in the improvement of battery design to provide more economical energy storage. The success of developing more economical storage batteries will have a significant effect on WEC economics.

Figure 7-4 lists some of the major considerations in utilizing electrochemical energy storage with WEC systems. One major problem is the cost generated at the interface to regulate the power input to the electrolyzer or battery. Arrangements must be made to control the power input since the output from the wind machine varies with the wind speed.

Potential Energy Storage Systems

The two primary potential energy systems are pumped hydro and compressed air storage. Pumped hydro applications have

**CONSTANT VOLTAGE PUMPING
& CONVERSION**
Concerns: (1) HIGH ENERGY PRODUCTION
            COSTS - $50-100/kW
(2) EFFICIENCY - ?
(3) CAPITAL COST - HIGH
(4) LIFE CYCLE - ?

BATTERIES
(1) EFFICIENCY - 70%
(2) EXPECTED COST:
  (a) LITHIUM - SODIUM
      $20-40
  (b) SODIUM - SULFUR
      $20-40

H_2/HYDRIDE/FUEL CELL
(1) EFFICIENCY - 50%
(2) EXPECTED COST:
    $30-90/kWH

Figure 7-4. Key considerations in electrochemical energy storage systems.

already been discussed. This is by no means a new idea. Storing energy by pumping water to high storage reservoirs for later use has been employed, although not to any great extent, for a number of years. The application of this type of energy storage arrangement has, however, grown in importance over the past several years. Table 7-1 illustrates the growth of pumped storage in applications to hydroelectric plants in terms of additional megawatts contributed. Additions of hydroelectric power plants are also compared with the total electrical generating capacity.

Compressed-air storage systems are another possible arrangement. Figure 7-5 illustrates a constant-pressure, near adiabatic compressed-air energy storage arrangement proposed by Zlotnik.[29] The actual storage compartment is referred to as the air receiver. Air receivers can be simple volume tanks. In general,

Table 7-1. Annual Hydroelectric Power Additions[30]

Year	Additional Power Capacity (MW)		
	Hydro	Pumped Storage	Total Generating Capacity
1969	2.3	0.8	24.1
1970	2.1	0.2	32.1
1971	1.1	1.1	32.6
1972	0.5	0.9	37.9

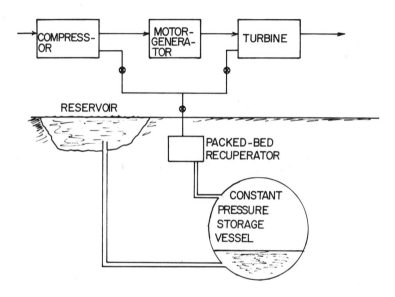

Figure 7-5. Constant pressure, near adiabatic scheme for compressed-air storage.

a volume tank is not often thought of as a highly engineered piece of equipment; however, if careful engineering is applied to receivers, significant reductions in equipment costs can be realized. The concept has been frequently employed in industry for a long time. Quite often there is the intermittent requirement for a large volume of air at moderate to high pressures to be stored for short intervals of time. An example of this are some boiler soot blowing systems. The purpose of these systems is to store air under pressure to supplement to compressor output when the demand arises. In general, the receiver volume required (*i.e.*, the cubic feet of air to be stored) represents the free air at 14.7 psia that must be packed into the vessel above the minimum pressure that is required by the demand. In sizing the receiver, allowances must be made of line losses. The final design can only be made after careful consideration of the cost of the compressor, the motor, starter, aftercooler, receiver, and the cost of installation which includes foundations, piping and wiring. In addition, power costs including demand charges must be considered. Such systems can amount to a considerable investment, and as such, much effort is needed in incorporating a WECS into an existing system. The proposed scheme shown in Figure 7-5 does not, for example, require an aftercooler on the compressor. Hence, for this arrangement the energy that is generated by the work of compression can be conserved in a recuperator (as established by a temperature equal to that of an adiabatic process). When operated in the storage mode, the turbine is decoupled and the compressor pressurizes the receiving chamber. When operated in the generation mode, the compressor is decoupled and the stored air is expanded through the turbine to drive the generator.

Spinoffs from this type of storage application are hydrogen gas storage and the production and storage of methane gas. The latter is more of a direct application of wind energy than a collection arrangement but is worthy of some discussion. Basically, one concept that has recently been suggested is the use of off-shore winds to obtain hydrogen through electrolysis of seawater. Carbonate deposits supply carbon dioxide, which, when combined with hydrogen, forms methane (CH_4). Recall that conventional wind machines employed in generating a.c. electricity extract only a small fraction of the energy available in the wind. The a.c. generators are driven at constant speed, and as wind speed increases the windmill blade pitch must be readjusted to maintain a constant blade rotational speed. Hence, the aerogenerator harnesses considerably less than the maximum power potential of the wind. It has been suggested that d.c. generators could supply current to run electrolysis cells. Electrolysis cells can function over a wide range of currents, and wind-driven machines could extend their range of power harnessing from the wind. Such an arrangement would allow the wind turbine to operate at the designed tipspeed for maximum power by adjusting the load, instead of the rotational speed fixed by the a.c. generator. In addition, it reduces the need for cut-in and cut-out speeds in a.c. networks as wind speeds vary.

The approach is a good idea, especially when one considers that the U.S. already has an enormous investment in natural gas equipment such as transmission lines and user equipment, as well as huge storage facilities. Methane is a major constituent in natural gas and thus a likely candidate as a natural gas substitute as these supplies are diminished. The conversion of wind energy to hydrogen and then to CH_4 readily establishes a large energy storage facility. There are several reasons why offshore installations have been suggested. As previously noted, wind velocities are generally higher and more consistent along shorelines and offshore. Also, water and carbonates, two of the necessary raw materials, are readily available.

Thermal Energy Storage

By heating, melting or vaporizing a material, energy can be stored. When the process is reversed, energy becomes available in the form of heat. Sensible-heat storage is a method of saving energy by elevating the temperature of a material. The efficiency of this approach relies on the specific heat of the material and the density of the storage material. In English units, the specific heat is the number of Btus required to raise 1 lb 1°F.

Another form of thermal storage is by phase change, *i.e.*, the transition from solid to liquid state or from liquid to vapor. This form is known as latent-heat storage, in which no change in temperature is involved.

It is possible for both sensible-heat and latent-heat storage to occur in the same material; for example, when a solid material is heated, then melted, and then further raised in temperature. The heat of fusion is usually of the same order of magnitude of the product of specific heat and the temperature change. Tables 7-2 and 7-3 list several materials suitable for sensible-heat and latent-heat storage, respectively.

110 FUNDAMENTALS OF WIND ENERGY

Table 7-2. Materials Suitable for Sensible-Heat Storage

Material	Specific Heat (Btu/lb °F)	Density (lb/ft³)	Heat Capacity (no voids) Btu/ft³ °F
Brick	0.20	140	28
Concrete	0.27	>140	38
Magnetite (Fe_3O_4)	0.18	320	57
Scrap Aluminum	0.23	170	39
Scrap Iron	0.12	490	59
Sodium*	0.23	59	14
Water	1.00	62	62

*Up to 208°F

Table 7-3. Materials Suitable for Latent-Heat Storage

Material	Melting Point (°F)	Density (lb/ft³)	Heat of Fusion (no voids) (Btu/lb)
Calcium Chloride Hexahydrate	84-103	103	75
Ferric Chloride	580	181	114
Hypophosphoric Acid	131	95	92
Lithium Hydride	1260	51	1800
Lithium Nitrate	480	150	158
Sodium Hydroxide	612	133	90
Sodium Metal	208	59	42

Water is exceptionally good for sensible-heat storage in temperature ranges below 200°F. Sensible-heat storage with air as the transport mechanism has also been suggested. This involves using crushed stone or gravel in a bin, which acts as a large and economical heat-transfer surface.

No suitable materials have been found for latent-heat storage near ambient conditions. There are great limitations on materials required for melting slightly above ambient conditions. Organic materials, for example, are usually comprised of weak crystal lattices so that their heat of fusion is low. Inorganic salts usually melt at higher temperatures (unless they are hydrates). An ideal material for latent-heat storage would be an inorganic hydrate where the water content of the melted and unmelted portions are the same. This would result in a material with a congruent melting point. The major problem associated with known hydrates is that they separate into an anhydrous solid residue and a dilute solution, *i.e.,* there is no true melting.

Magnetite (iron ore) is capable of storing heat over a wide temperature range. Interestingly, it exhibits a volumetric heat capacity in a temperature range up to about 1000°F, approximately the same as water at much lower temperatures. Concrete is not as good as magnetite over a number of years to supplement home-heating loads.

Thermal storage systems are still very much in the developmental stage. It is not yet clear whether they will be commercially available by 1980 to meet a major WEC program.

Mechanical Energy Storage

Storing energy in a flywheel is another possible storage arrangement for WEC units. This method of energy storage has also been suggested for operating an automobile. Since the last stages of steam turbines operate at top speeds of 2000 ft/sec, it is feasible that an automobile flywheel could be designed for 1000 ft/sec rim speed. It has been estimated that at this speed, the energy storage capacity of a flywheel could match the automobile storage battery.

There are a variety of designs of flywheels, each having certain limitations. Figure 7-6 indicates the range of energy density storage capacity for flywheel arrangements of different construction materials. In general, the energy density is proportional to the ratio of the flywheel's working stress to its material density. For known materials of construction, the range of variation is attributed primarily to the life of the flywheel. The lifetime is a function of the ratio of the working stress to the ultimate stress and also depends on the ratio of the maximum to minimum rotational speed. There have been no thorough economic studies on incorporating flywheel energy storage units in a WEC design. Further research into this area is necessary.

112 FUNDAMENTALS OF WIND ENERGY

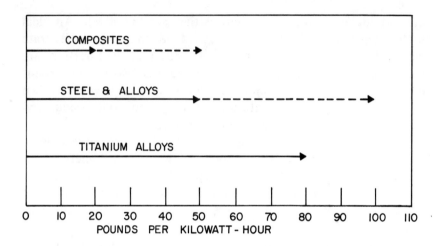

Figure 7-6. Range of energy density values for flywheels of various construction materials.

Economics of Storage Systems

There have been a sufficient number of economic studies of energy storage systems to allow comparison among the various storage technologies. Of particular interest are:

(1) the minimum economic sizes of storage systems for a particular utility application;
(2) the total capital investment for the entire energy conversion system;
(3) the capital investment for the energy-storage arrangements;
(4) the expected wind energy conversion system lifetime;
(5) the expected energy storage system lifetime; and
(6) the expected turnaround efficiencies of the storage system as well as the entire conversion unit.

Figure 7-7 indicates the estimated capital costs for various energy storage units within a WECS in terms of dollars per rated kilowatts for the minimum economic size electric utility. The total capital cost comparison shown is based on an energy storage cycle of approximately 10 hr. These estimates are not highly reliable since this may not be nearly enough time for many wind base load applications. Under actual operation, storage times may

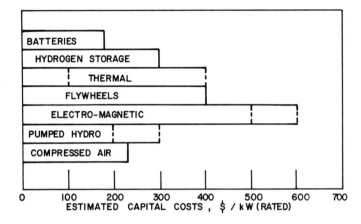

Figure 7-7. Estimated capital costs of various energy technology systems.[30]

be on the order of several days in order to compensate for short-time wind speed variations, or storage cycles may very well last 6 months to a year depending on seasonal changes in wind climatology. This brings up a further point, although it is rather obvious: the final design and selection of a particular storage system cannot be based on the economics alone. Other aspects must be examined. Table 7-4 compares the expected life cycles of the various storage arrangements and their turnaround efficiency.

Table 7-4. Comparison of Expected Life Cycle of Various Energy Storage Systems[31]

	Minimum Economic Sizes for Utility Usage (MWh)	Expected Lifetime (Yr)	Turnaround Efficiency (%)
Batteries	10	10-20	70-80
Hydrogen Storage	10	30	40-60
Thermal	600	20	?
Flywheel	10	30	80
Electromagnetic	10^4	30	90-95
Pumped Hydro	10^4	50	70
Compressed Air	100	20	45

114 FUNDAMENTALS OF WIND ENERGY

Estimates on the turnaround efficiency of thermal storage systems is now known; however, it is expected to be high. A further consideration is the adaptability of each storage system to different applications. For example, if long storage periods are required, there may be significant cost advantages in using hydrogen or compressed-air storage, especially if depleted natural gas wells or other natural underground caverns are already available for storing wind-generated energy.

Storage systems may also have to adjust to the physical layout of the wind conversion scheme. For example, in some applications, several small wind machines may have to be spaced apart: for example, in offshore methane production. Under such circumstances, the type of storage system used might have to be capable of being dispersed in the same fashion as the group of wind machines. Batteries, hydrogen storage, thermal storage and flywheel energy storage systems are capable of dispersed storage capabilities; however, electromagnetic, pumped hydro and compressed-air storage generally are not.

Finally, the impact that energy storage systems have on the consumer's economic outlook is of major importance. Some idea of this can be obtained by comparing the incremental increase in the busbar price of electricity for various energy storage systems. This can be estimated from the following equation:

$$I = \frac{CF}{(8.76) E L} + A \qquad (7.1)$$

where: I = the incremental increase in the busbar price for energy storage (mills/kWh),
C = the capital cost of the storage system ($ per rated kW),
F = the fixed charge rate,
E = the turnaround efficiency of the storage system (%100),
L = the load factor of the storage system, and
A = the operations and maintenance costs associated with the storage system (mills/kWh).

The reader can refer to the Glossary for more detailed definitions.

ENERGY STORAGE SYSTEMS 115

ENVIRONMENTAL AND SAFETY CONSIDERATIONS

There are a number of safety and environmental effects which must be carefully evaluated before commercializing WEC systems on a large scale. This section identifies the principal environmental and safety issues of WECS. Most of the effects on the environment are not well understood, or even known. To understand potential problems, the wind engineer must:

(1) conduct a thorough literature search and compile all information related to wind effects, alteration of wind climatology and other related areas of environmental effects of wind energy production;
(2) make initial short-term observations at suitable sites, in that field studies are badly needed;
(3) conduct long-term field programs to assess potential impacts which might develop over a course of several years and would not be detected in short-term observations.

Environmental effects may vary with the particular WEC program application. It is essential that all existing information be carefully organized and sorted according to the particular situation. Some of this information will include microhabitat locations and/or adaptations to wind streams, the effects of surface regimes on natural and/or agricultural systems, migratory pathways and possibly navigational dangers to migratory biota. Both ecological and meteorological data relative to the particular sites of interest should be acquired.

One area in which little information exists concerns the effect the wind turbine will have on the turbulent nature of the wind stream. To assess this concern accurately, careful evaluation of the zone of influence of the wind turbine must be made (refer to Figure 7-8). That is, both the vertical and the horizontal extent of the wind machine's effects on the surrounding physical environment must be known. Careful aerodynamic analyses must be conducted to determine what effects, if any, air disturbances will have on birds, insects, etc., within the rotor's zone of influence. Possible effects might include entrainment of these flying organisms during the wind turbine's operational wind speeds, or increased collision probability, or perhaps limited maneuverability. These possibilities suggest that ecological studies must be

116 FUNDAMENTALS OF WIND ENERGY

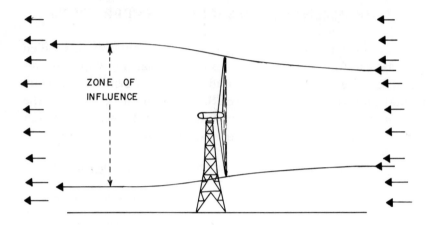

Figure 7-8. Careful evaluation must be made of the wind machine's zone of influence on a wind stream.

incorporated into site selection methodology. Detailed information on animal populations adjacent to a proposed facility, migratory and breeding bird surveys and information dealing with the patterns of migration or involuntary dispersal of insects as well as meteorological data might have to be acquired before selecting a final site. Determination of bird kill will also have to be considered for larger installations. Furthermore, little is known as to how this zone of influence will affect seasonal or daily wind climatology. Hence, long-term field studies are necessary to evaluate the variation of air temperature and wind velocity with height in the zone. Observations of effects such as evaporation, frost or fog occurrence will have to be made, as well as variations in surface temperature, nocturnal temperature inversions and precipitation. If these effects are indeed significant, detailed information on plant as well as animal species and populations will have to be incorporated in site selection methodology. Once these potential impacts have been carefully identified, careful siting and/or design alterations may eliminate or at least minimize such adverse effects.

To complicate matters further, the question arises as to what effect group arrangements have on the environment. Even less information exists on this topic. It is certainly of great interest to learn what effect groupings may have on the energy flow in the atmospheric boundary layer. Wind machines can be thought of as energy sinks with respect to the boundary layer, and, as indicated above, may greatly alter the wind profile.

One further question that must be addressed is how rapidly will energy be replenished in the wake of a windmill? The answers to this question will aid in determining the optimum distances between the various units in a WEC battery. It is quite possible that the energy removed by one wind machine can be replenished by the turbulent energy flow aloft or from adjacent arrangements. Volumetric air-flow rate is dependent upon vertical wind, temperature and density distributions in the boundary layer, and these may undergo significant alterations by the presence of batteries of wind machines. The structure of the turbulence wake and its extent depend on the structure and turbulence generated by each mill; furthermore, each unit may display different characteristics. Considerably more detailed information is required on wind and temperature distributions in the boundary layer before a full assessment of these phenomena can be made.

Because of the lack of detailed quantitative information on the adverse environmental effects identified above, there is considerable controversy over the degree of impact from wind machines. Some experts feel that climatic disturbances generated by extracting energy from the wind can be no more great than what we now tolerate from tall trees or skyscrapers.[32] In comparison with hydro, nuclear and fossil energy sources, many consider wind energy completely clean.

Physical Disturbances

Horizontal- and vertical-axis turbines can create physical disturbances such as electromagnetic wave interference and obstructions to fauna.

Wave interference can arise when two or more waves having the same frequency melt at a receiver point and combine. The resulting wave that forms has an amplitude and phase equal to the sum of the two interfering waves. Stationary wind machines may generate such disturbances, which will affect radios, television reception, microwaves and satellite communications. Reflections from the rotating blades of a wind turbine can cause modulations of television signals. These disturbances can be significantly more noticeable than stationary multipath images that are sometimes produced by signal reflections from stationary structures such as tall buildings, or towers. The effects produced by rotating blades fortunately tend to diminish several hundred yards from the rotor. Wind machines will have to be subject to FCC standards (Federal Communications Commission) to minimize this problem, both for television and radio reception. These standards stipulate that the ratio of the primary signal (*i.e.*, the intended signal) to the secondary signal level (reflections from wind turbines) should not be less than 46 decibels (db).

The second problem, obstruction to fauna, has already been discussed. Very limited studies have shown that there is roughly a 1% probability of a bird or insect being struck by rotating blades assuming no action is taken to avoid the rotor. These studies are not conclusive, however, and further observations may be necessary to obtain data on bird kill. Vortex generators are not a problem in this area since most proposed designs are enclosed and turbines are located at the base of the tower. One possible solution to fauna obstruction is the use of artificial sound or fluorescent paint to frighten away winged organisms.

WEC systems create other types of physical disturbances. Recall that there are numerous components in a WECS, such as gear trains, electric generators, pumps, compressors, energy storage facilities, etc., and a significant amount of noise and thermal pollution is produced.

Noise levels generated by WEC units are expected to vary with the system design. Certain designs, such as the vortex generator which has noise levels in the 70 to 100 dbA range, may exceed occupational noise standards. Hence, special noise reduction construction materials or insulation may have to be used in the design. Table 7-5 lists permitted noise exposure limits as outlined by the Walsh-Healey Act for working environments.

The limiting daily exposure times for nonoccupational noise conditions are considerably more stringent. The environmental performance criteria for industrial facilties is less than 90 dbA, whereas ambient exposure standards require less than 50 dbA.

Table 7-5. Noise Tolerance Levels for Occupational Exposures

Sound Level (dbA)	115	110	105	100	95	90
Allowable Limiting Daily Exposure Time (hr)	0.25	0.5	1.0	2.0	4.0	8.0

One further physical hazard associated with the vortex generator is the possibility of tornado generation under certain conditions. Tornadoes produced by nature are caused by Coriolis forces that generate counterclockwise circulation of air currents around low-pressure regions in the northern hemisphere and clockwise circulation in areas in the southern hemisphere. This is one potential hazard that will have to be examined in great detail. One possibility is that vortex generators may be designed to circulate the wind in the opposite direction from natural tornadoes. Coriolis forces would tend to dampen the circulating wind and counteract this effect.

Air Pollution Problems

In general, no combustion of fossil fuels takes place in generating electric energy by a wind energy conversion system. For the most part, then, WEC systems are considered clean systems with respect to air quality control. However, energy storage units can be a source of air pollution. For example, batteries or electrolysis units can emit dangerous acid vapors. Because WEC units are limited in use at present, there are minimal chemical pollutant releases. If, however, large-scale programs do become a reality in this country, ambient and/or emission standards requirements will have to be formulated. This may mean the restriction of implementing WEC systems in residential areas and certainly would greatly affect the overall economics if air pollution control systems must be incorporated into designs.

Occupational Safety Problems

In addition to possible noise pollution problems, WEC systems using horizontal or vertical-axis turbines may create certain other

120 FUNDAMENTALS OF WIND ENERGY

occupational health and safety problems. Potential hazards include:

    (a) obstruction or danger to aviation traffic control,
    (b) disruption of navigation by offshore installations,
    (c) danger to people in the vicinity because of possible stress failure of WECS components such as blades, tower structure, etc.

Establishing WEC plants near or around airports is a key issue. For large MW-scale electric utilities, the size of blades, towers, buildings and other WECS components is such that they can significantly obstruct aviation. To meet MW-scale requirements, towers and blades of horizontal or vertical-axis turbines may have to be up to 100 m in height and could very easily obstruct takeoff and landing flight patterns. In addition, these units could easily interfere with flight control radar or radio communications by generating electromagnetic wave interferences. Wind engineers will have to work closely with the FAA when considering potential wind energy sites near airports. Units will have to be clearly marked on flight maps and suitable lighting arrangements made for night flights.

Other precautions will have to be instituted for offshore installations. If units are positioned in coastal or inland waters, WECS components may present obstructions to navigation. Safety zones several hundred feet in diameter may be required to prevent problems. In addition, systems will have to be properly equipped with lighting and foghorns for night. Here, the wind engineer will have to work closely with the Coast Guard in selecting potential sites and in locating installations on U.S. coast and geodetic survey charts. Figure 7-9 illustrates what a typical WECS offshore installation looks like.

The greatest potential for safety problems may arise from turbine blade failure caused by various stresses imposed on the unit. Tower support stability is a major engineering consideration. In general, high structures such as towers are engineered to meet minimum wind loading. Wind machines, on the other hand, must be designed to meet maximum wind loading at the top of the tower. Even for small-scale systems, the lateral force near the top can be significant to performance. The blades of these large

ENERGY STORAGE SYSTEMS 121

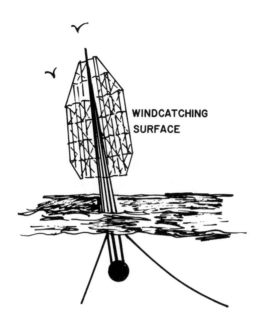

Figure 7-9. Typical offshore WECS arrangement.

rotating machines may be tens of meters in length, and it is important that their rotation frequency does not match any of the structural resonances of the support towers. (Note that if the rotation speed is permitted to vary with wind speed, interference with structural resonances cannot be avoided.)

Gravitational forces, tower shadowing, wind gusts, wind shear, etc., impose large stresses on support towers and particularly on the blades. Prolonged or excessive stresses can cause blades to break. Carefully engineered systems will have to meet standards set by the Occupational Safety and Health Act (OSHA). It is likely that additions for WEC systems will have to be made to the general industrial standards outlined by OSHA. For example, criteria may have to be established for minimum height of blade tips above the ground for wind turbines.

A further problem that may arise is the operation of wind machines in cold climates. Ice accumulation could become excessive and lead to blade failure. Furthermore, ice shedding from blades is a potential safety hazard.

CLOSING REMARKS

It has been the goal of this chapter to identify and assess some of the environmental impacts and/or effects of the operation of wind energy conversion systems. It should be clear that further investigation into these areas is necessary before a large-scale WECS can be implemented. Although little attention has been given to environmental impacts from energy storage systems, the reader should be made aware that these may very well become the chief offenders in health, safety and environmental violations. Figure 7-10 summarizes the key environmental issues discussed above.

- E.M. WAVE INTERFERENCE
- OBSTRUCTIONS TO FAUNA
- POTENTIAL TORNADOS
- NOISE
- LAND-USE PROBLEMS
- OFFSHORE USE PROBLEMS
- AESTHETICS
- HAZARDS TO AVIATION
- HAZARDS TO NAVIGATION
- HAZARDS FROM ROTATING BLADES & COMPONENTS

Figure 7-10. Summary of potential environmental impacts from WECS.

CHAPTER 8

FUTURE POTENTIAL OF WIND ENERGY

"Steam is no stronger now than it was a hundred years ago, but it is put to better use." . . Emerson

PUBLIC ACCEPTANCE OF WIND ENERGY

Different types of wind energy applications and designs may generate varied public opinion or varying degrees of socioeconomic disturbances. If WEC programs are to play a significant role in our future energy requirements, the reactions of the general public across the country toward the construction and operation of wind machines must be explored. There are essentially three types of socioeconomic considerations which will have to be contended with. These are:

(1) land-use requirements for installations,
(2) offshore area requirements, and
(3) aesthetics.

Typical land-use requirements for a WEC plant are estimated to be on the order of 0.25-0.50 ac/MW of electricity produced. This is in reality quite small when compared to land-use requirements for solar and some of the conventional fuel systems. Turbines are mounted either on or in towers, thus freeing the land surrounding the installation for other purposes. Multiple units or groupings of wind machines may be more difficult to sell to the public, as dispersed MW-scale systems need far more land. Transmission lines, for example, can take up a large portion of these additional requirements.

At present there are no national standards that restrict land-use requirements for WEC arrangements; therefore, local and state zoning laws will have to be relied on heavily. Furthermore, safety perimeters and zones will have to be incorporated into land-use requirements for a system.

Offshore installations are appealing because of the availability of high and persistent winds. Systems can be constructed on the continental shelf, large lakes, etc. Major public criticism may arise because of obstructions to the navigation of small craft in recreational areas. Usually, offshore installations are mounted on either floating or stationary platforms. In addition, large systems may be unappealing from an aesthetic viewpoint, particularly around scenic or seashore recreational areas. Potential restrictions may include limits on the number of offshore installations per unit area, and this in turn will rely on the original intended purpose for the area and public reactions. If public opinion is strongly against these systems in general, offshore units may be allowed to be constructed only in remote regions or outside shipping lanes.

The aesthetics of wind machines may be one of the most crucial constraints on the widespread usage of WECS. There are many examples of technological developments that are economical and practical but, for reasons of public dissatisfaction, had to be either discarded or drastically modified. It is therefore imperative that aesthetic considerations play a significant role in the ultimate design of a wind machine for a particular application.

LARGE-SCALE WIND POWER

A number of potential applications have been discussed; among the most promising on a large scale are electricity generation and fuel production.

Most wind energy experts agree that WEC programs would be capable of providing electricity into a grid. Such systems could be employed as fuel savers or, in situations where energy storage is available, as a supply of intermittent, supplementary or peaking power. Also, independent industries with large electricity demands could use their own WECS without having to connect with a utility grid. Various studies have indicated that WEC systems could supply 5 to 15% of the present U.S. electricity requirements.

No one will argue that there is an urgent need to develop alternate sources for clean, competitive energy as quickly as possible. Rapid increases have taken place in the cost for energy obtained from conventional energy supplies and systems. The technology is rapidly approaching a level that will make WEC systems viable on a large-scale commercial basis, despite the large number of unknown factors that were identified in Chapter 7. Indeed, there are no major problems in constructing wind machines. In general, wind turbines are considerably less complex than aircraft. Furthermore, wind surveys, though not entirely reliable because of the lack of methodology, seem to indicate a sufficient number of wind energy sites in the United States.

AREAS OF RESEARCH

It is the opinion of the author and of many energy experts that WEC systems offer mankind a practicable alternative source of energy. Many worthwhile research programs have been initiated in the early half of this decade by the National Science Foundation (NSF) and the Energy Research and Development Agency (ERDA) on wind energy related topics. Federal support is being provided to accelerate the design, construction and operation of large-scale WEC systems.

A high-priority research area, and one whose outcome will be cost-effective, is the development of methodology and systematic approaches to locating and evaluating wind energy sites. Incorporated into these studies must be the development of more appropriate instrumentation for measuring and analyzing wind climatology. At present, measuring equipment is costly and does not provide reliable quantitative data for the wind engineer.

Currently, the wind engineer is relying on the national meteorological service and a number of government organizations for wind climatology information. In general, they are well organized at archiving data; however, the information they supply can at best be used only for initial screening of potential wind power sites. There is an urgent need for long-term field studies to help us understand the physical parameters involved in the turbulent atmosphere around the earth. Armed with better instrumentation and a more physical understanding of wind energy related parameters, the wind engineer can develop adequate mathematical models to describe the atmospheric boundary layer. Once this has been accomplished, methodological wind survey programs can be initiated.

It should be emphasized that the ultimate decision to develop large-scale WEC production will be based on the power output per unit capital cost and not on the efficiency of such systems alone. The economics of WEC systems must be made competitive with the conventional alternatives. This may mean that improvements will have to be made on aerodynamic and electrical designs in order to harness a greater portion of the wind's power. At the same time, engineers will have to work toward less costly designs and construction materials. The most elaborate horizontal-axis wind machines are already capable of extracting over half the energy in the wind at the most favorable wind speeds, yet there is still a great deal of room for improvement, and research is needed for reducing costs through better engineering. In order for a volume market for wind-operated energy-generating systems to develop in this country, systems must first be proven economically feasible. This would provide incentive to manufacturers to invest in tooling and design improvements in order to minimize costs.

To quote an old English proverb, "Time and tide wait for no man." The need for new energy sources is now, and we must quickly assess the role of wind energy in the present as well as in the near and distant future.

The authors believe that a successful discussion should raise new and challenging questions. What better way is there to end our discussion of wind energy than by suggesting questions that should be answered by today's wind engineer:

• Will energy storage facilities be required for all WEC systems, and, if so, what impact will this have on the cost-effectiveness of such systems?

• What types of energy products are needed for different WEC applications?

• Will elaborate energy transmission systems be required for connecting dispersed WEC systems to utility networks?

FUTURE POTENTIAL OF WIND ENERGY

- What are the primary impediments standing in the way of implementing large-scale WEC systems?
- What specific engineering is necessary to remove these impediments while at the same time minimizing costs?
- How large should large-scale wind machines be, and what are the exact design criteria for size?
- What significant impacts will WEC systems have on environmental, safety and health aspects?
- Should a cooperative international program be organized to develop WEC?
- What will be the eventual cost per kWh of electricity produced for different large-scale utility arrangements?

APPENDIX A

GLOSSARY OF ENERGY-RELATED TERMS

a.c.: Alternating current.

Air pollution: The presence of unwanted or harmful material in the atmosphere. Unwanted material refers to components in sufficient concentrations, present for a sufficient time and under certain circumstances to interfere significantly with the comfort, health or welfare of people and the environment.

Aquifer: A water-bearing stratum of permeable rock, sand and/or gravel employed for storage of gas.

Average wind speed: Mean wind speed over a specified length of time.

Bedplate: A base plate used for supporting a structure.

Boiler: A closed pressure vessel in which a liquid is vaporized by the application of heat.

Boiling: The conversion of a liquid into its vapor state.

Breakeven costs: The system costs where the price of the system's product matches the cost of the equivalent energy product of a different type of system.

British thermal unit (Btu): The mean British thermal unit is 1/180 of the heat needed to raise the temperature of one pound of water from $32°F$ to $212°F$ at a constant atmospheric pressure. (1 Btu $\simeq$ 252 calories; 3413 Btu = 1 kWh.)

Busbar price: The price of electricity at a generating plant excluding price increments resulting from distribution and transmission costs.

Capacity factor:	The ratio of the average load on a machine for the period of time considered to the capacity rating of the machine or equipment.
Capital costs:	Investment costs necessary to construct a system.
Chord:	The length or distance from the leading to the trailing edge of the foil.
Coal gas:	Gas formed from the destructive distillation of coal.
Coal tar:	Black viscous liquor, one of the by-products formed from the distillation of coal.
Coke:	Fuel consisting primarily of fixed carbon and ash in coal obtained by the destructive distillation of bituminous coal.
Coke oven gas:	Gas generated by destructive distillation of bituminous coal in closed chambers (heating value approximately 550 Btu/ft^3).
Coking:	The conversion of a carbonaceous fuel by heating in the absence of air.
Combustion:	The chemical combination of oxygen with the combustible elements of a fuel, resulting in the generation of heat.
Concentrator:	A device that concentrates a wind stream.
Conventional fuels:	The fossil fuels: coal, oil and gas.
Cracking:	Thermal decomposition of complex hydrocarbons into simpler compounds or elements.
Crosswind:	Crosswise to the direction of the wind stream.
Crude oil:	Unrefined petroleum.
Cut-in speed:	The wind speed at which an aerogenerator is activated as the wind speed increases.
Cut-off speed:	The wind speed at which a wind machine is designed to shut off to prevent damage from high winds.
d.c.	Direct current.
Demand:	The rate at which electric energy is delivered to a system, part of a system or a specific piece of equipment, usually expressed in kilowatts, kilovolt-amperes or other suitable units at a given instant or averaged over a specified period of time; the power-consuming equipment of the customers.

Annual Maximum. For a particular load under consideration, the largest of all demands of the load which occurred during a prescribed demand interval in a calendar year.

Average. The power output of an electric system or any portion of its components over any time interval, as evaluated by dividing the total number of kWh by the number of units of time in the interval.

Billing. The demand upon which billing to a customer is based. It may be based on the contract year, a contract minimum, etc., and hence does not always coincide with the actual measured demand during the billing period.

Instantaneous Peak. Maximum demand at the instant of greatest load, generally determined from readings from indicating meters.

Maximum. The largest of all the demands of the specific load considered during a specified time period.

Depreciation: With reference to a depreciable electric generating plant, the loss in service value not restored by maintenance, incurred in connection with the consumption of the electric plant in the service from causes which are known to be in the current operation and against which the utility is not protected by insurance. Wear, decay, inadequacy, obsolescence, changes in technology and/or changes in demand of public authorities are among the causes to be examined.

Diffuser: A device that spreads out or diffuses a wind stream.

Distribution: The method of transporting or distributing electric energy from convenient locations in the transmission system to the consumers.

DOE: Department of Energy—a federal cabinet-level department created in 1977.

Downwind: On the opposite side from the direction of the blowing wind.

Draft tube: A flared passage leading vertically from a water turbine to its tailrace.

Drag-type systems: Devices that are actuated by aerodynamic drag in a wind stream (*e.g.,* Savonius rotor).

Dry fuel basis:	The method of reporting fuel analysis without moisture or other constituents.
ECS:	Environmental control systems.
Energy, Electric:	As used in the electric utility industry, electric energy is kilowatt hours (kWh).
ERDA:	Energy Research and Development Administration, superseded by the Department of Energy (DOE).
Eutectic salt:	A salt compound having a large heat of fusion and a low melting point.
Fuel:	A substance that contains combustible materials used for generating heat.
Fuel oil:	A liquid fuel derived from coal or petroleum.
Generating station:	A station or plant at which are located prime movers, electric generators and auxiliary equipment for converting mechanical, chemical or nuclear energy into electric energy.
Generating unit:	An electric generator together with its prime mover.
Generation, Electric:	Refers to the act of transforming other forms of energy into electric energy; the amount of electric energy so produced, expressed in kWh. *Gross:* Total amount of electric energy produced in a generating station. *Net:* Gross generation minus the kWh consumed out of gross generation for station use.
Generator, Electric:	A machine that transforms mechanical energy into electrical.
Generator, Steam:	Machinery which burns fuel and transforms water into steam.
Grain:	Unit of weight (1 pound = 7000 grains).
Gross National Product (GNP):	The sum of personal consumption, expenditure of goods and services, plus government expenditure on goods and services, plus investment expenditure, in dollars. The GNP gross investment expenditure on all new machines and construction is included.

Headrace:	A channel leading to a water turbine.
Heat exchanger:	A device that transfers heat from one fluid to another.
Heat rate:	A measure of a generating station's thermal efficiency, usually expressed as Btu per net kWh and calculated by dividing the total Btu content of fuel burned or consumed for electric generation by the resulting net kWh generation.
Heat sink:	Anything that absorbs heat (*e.g.,* air, a lake).
hp:	Horsepower—a measure of power capacity.
Interchange:	Kilowatt-hours delivered to one electric utility system by another for economic reasons. This power may be returned at a later time or accumulated as energy balance until the end of an agreed-upon interval.
Ion:	A charged atom or radical group having either positive or negative charge.
Ionization:	The process of adding electrons to or removing them from atoms or molecules, thus making ions. Various causes of ionization are high temperatures, electrical discharges and nuclear radiation.
kVa:	Kilovolt-amperes (10^3-volt amperes)—a measure of power capacity.
kW:	Kilowatt (10^3 watts)—a measure of power.
kWh:	Kilowatt-hour (10^3 watt-hours)—a measure of energy.
Lift-type devices:	Systems that employ air-foils or other devices to provide aerodynamic lift in a wind stream.
Liquified Petroleum Gas (LPG):	Comprised of hydrocarbons that are usually gases at normal atmospheric conditions but can be liquified under moderate pressures. LPG is derived from natural gas and various refinery sources (*e.g.,* crude distillation and cracking).
Load:	The amount of electric power delivered at any specified point or points in a system. The primary load originates at the power-consuming equipment of customers.
Load factor:	Average power output of an energy system divided by its rated power output.

134 FUNDAMENTALS OF WIND ENERGY

Megawatt (MW): 10^3 kilowatts (or 10^6 watts)—a measure of power.

mph: Miles per hour.

Name plate rating: The full-load continuous rating of a generator and its prime mover normally indicated on a name plate attached mechanically to the machine. The name plate rating is usually less (for older equipment and generators, sometimes more) than the demonstrated capability of the installed generator.

Natural gas: Gaseous fuel occurring in nature, principally methane.

Net capability: Gross station output without the portion used for station services such as auxiliaries.

NASA: National Aeronautics and Space Administration.

NCC: National Climatic Center.

Net for distribution: On an electrical system basis, this refers to the kWh available for the total system or company load. It is the sum of the net generation by the system's own plants, the purchased energy and the net interchange.

NOAA: National Oceanic and Atmospheric Administration.

Off-peak: Energy supplied during periods of low system demands, as specified by the supplier.

Panemone: Vertical-axis wind machine capable of reacting to horizontal winds from any direction.

Peak-shaving: The method of supplying power from an extraneous source to aid in meeting the peak demand on a system.

Penstock: Conduit for conveying water to a power plant.

Petroleum: Naturally occurring mineral oil composed predominantly of hydrocarbons.

Pinned truss: An assembly of connected beams.

Polder: A tract of low land reclaimed from the sea by dikes or dams.

Power: The rate at which energy is consumed. [English units: kilowatts (kW), megawatts (MW) or horsepower (hp); 1 MW = 10^3 kW; 1 hp = 0.746 kW.]

Power coefficient: Ratio of the power extracted by the windmill to the power available in the wind stream.

GLOSSARY 135

Power density: Amount of power per unit of cross-sectional area of wind stream.

Pressurized water reactor: Power reactor where heat is transferred from the core to a heat exchanger by water kept under high pressure to achieve high temperature and deter boiling from occurring in the core. Steam generation occurs in the secondary circuit.

Prime mover: The engine, turbine, water wheel, wind machine or similar devices which drive an electric generator.

Production: The act of generating electric energy; functional classification referring to the section of the utility plant that is used for generating electricity; also refers to expenses related to the operation or maintenance of a production plant and to the purchase and interchange of electricity.

Pumped storage: A method that fills a reservoir with water by pumping during off-peak periods when low-cost steam energy is available or when water is being spilled at neighboring hydro plants. The stored water can be used at more appropriate times.

Rated output capacity: Output power of an aerogenerator that is operating at the constant speed and output corresponding to the rated wind speed.

Rated wind speed: The lowest wind speed for which the rated power output of a wind collector is manufactured.

REA: Rural Electrification Administration.

Refinery gas: Commercial noncondensible gas derived from fractional distillation of crude oil or cracking of crude oil on petroleum distillates. Refinery gas is usually burned at the refinery or sold to public utilities for mixing with city gas.

rpm: Revolutions per minute.

Scroll: A conical casing which directs water to the turbine.

Shear: A relative force parallel to the surface of contact.

Shroud: A device employed to concentrate or deflect a wind stream.

136 FUNDAMENTALS OF WIND ENERGY

Smock mill: A 17th-century Dutch windmill resembling an artist's smock.

Station use,
 Generating: The power (kWh) used at an electric generating station for purposes such as excitation and operation of auxiliary and other facilities necessary for the operation of the plant. The station use includes the electric energy supplied from house generators, main generators, transmission networks and other sources. The quantity of energy used is the difference between the gross generation and the net output of the station.

Steam: Vapor phase of water significantly unmixed with other gases.

Steam-electric
 generating plant: An electric generating station that utilizes steam for the motive force of its prime movers.

Stock: A bar employed as a support for a windmill sail.

Superheated steam: Steam that is at a higher temperature than its saturation temperature.

Tailrace: Channel leading away from a water turbine or similar device.

Tip-speed-to-wind-
 speed ratio: The ratio of the speed of the tip of a propeller blade to the speed of the wind stream in which it lies.

Tjasker: A 14th-century Dutch windmill which housed an Archimedean screw and was used for pumping water.

Translational motion: Motion in a straight line (vector).

Transmission: The method of transporting electric energy in bulk from a source in the generating system to other utility systems.

Transmission line: A line employed for bulk transmission of electrical energy between a generating or receiving point and delivery stations or points.

Turbine generator: A rotary-type system consisting of a turbine and an electric generator.

Turbine, Steam or
 Gas: Enclosed rotary class of prime mover where heat energy in steam or gas is converted into mechanical

GLOSSARY

	energy by the force of a high-velocity flow of steam or gases directed against successive rows of radial blades that are fastened to a central shaft.
Turnaround efficiency:	The resulting efficiency when energy is transformed from one state to another and then reconverted to its original state.
Upwind:	On the same side as the direction from which the wind is blowing (in the path of the oncoming wind stream).
WEC:	Wind energy conversion.
WECS:	Wind energy conversion system.
Wind rose:	Pattern formed by a diagram illustrating vectors representing wind velocities.
Wind turbine generator:	A wind machine that drives an electric generator.

APPENDIX B

CONVERSION FACTORS

Multiply	By	To Obtain
Acres	43.560	Square feet
Acres	4,047	Square meters
Acres	1.562×10^{-3}	Square miles
Acres	4840	Square yards
Acre-feet	43.560	Cubic-feet
Acre-feet	3.259×10^5	Gallons
Angstrom units	3.937×10^{-9}	Inches
Atmospheres	76.0	Centimeters of mercury
Atmospheres	29.92	Inches of mercury
Atmospheres	33.90	Feet of water
Atmospheres	10,333	Kilograms/square meter
Atmospheres	14.70	Pounds/square inch
Atmospheres	1.058	Tons/square foot
Barrels (British, dry)	5.780	Cubic feet
Barrels (British, dry)	0.1637	Cubic meters
Barrels (British, dry)	36	Gallons (British)
Barrels, cement	170.6	Kilograms
Barrels, cement	376	Pounds of cement
Barrels, oil	42	Gallons (U.S.)
Barrels, (U.S., liquid)	4.211	Cubic feet
Barrels, (U.S., liquid)	0.1192	Cubic meters
Barrels (U.S., liquid)	31.5	Gallons (U.S.)
Bars	0.9869	Atmospheres
Bars	1×10^6	Dynes/square centimeter
Bars	1.020×10^4	Kilograms/square meter
Bars	2.089×10^3	Pounds/square foot
Bars	14.50	Pounds/square inch

140 FUNDAMENTALS OF WIND ENERGY

Multiply	By	To Obtain
Board-feet	144 square inches x 1 inch	Cubic inches
British thermal units	0.2520	Kilogram-calories
British thermal units	777.5	Foot-pounds
British thermal units	3.927×10^{-4}	Horsepower-hours
British thermal units	1054	Joules
British thermal units	107.5	Kilogram-meters
British thermal units	2.928×10^{-4}	Kilowatt-hours
Btu (mean)	251.98	Calories, gram (mean)
Btu (mean)	0.55556	Centigrade heat units
Btu (mean)	6.876×10^{-5}	Pounds of carbon to CO_2
Btu/minute	12.96	Foot-pounds/second
Btu/minute	0.02356	Horsepower
Btu/minute	0.01757	Kilowatts
Btu/minute	17.57	Watts
Btu/square foot/minute	0.1220	Watts/square inch
Btu (mean)/hour $(ft^2)°F$	4.882	Kilogram-calorie/$(m^2)°C$
Btu (mean)/hour $(ft^2)°F$	1.3562×10^{-4}	Gram-calorie/second $(cm^2)°C$
Btu (mean)/hour $(ft^2)°F$	3.94×10^{-4}	Horsepower/$(ft^2)°F$
Btu (mean)/hour $(ft^2)°F$	5.682×10^{-4}	Watts/$(cm^2)°C$
Btu (mean)/hour $(ft^2)°F$	2.035×10^{-3}	Watts/$(in.^2)°C$
Btu (mean)/pound/°F	1	Calories, gram/gram/°C
Bushels	1.244	Cubic feet
Bushels	2150	Cubic inches
Bushels	0.03524	Cubic meters
Bushels	4	Pecks
Bushels	64	Pints (dry)
Bushels	32	Quarts (dry)
Calories, gram (mean)	3.9685×10^{-3}	Btu (mean)
Calories, gram (mean)	0.001469	Cubic feet-atmospheres
Calories, gram (mean)	3.0874	Foot-pounds
Calories, gram (mean)	0.0011628	Watt-hours
Calories, (thermochemical)	0.999346	Calories (int. steam tables)

CONVERSION FACTORS 141

Multiply	By	To Obtain
Calories, gram (mean)/gram	1.8	Btu (mean)/pound
Centigrams	0.01	Grams
Centiliters	0.01	Liters
Centimeters	0.0328083	Feet (U.S.)
Centimeters	0.3937	Inches
Centimeters	0.01	Meters
Centimeters	393.7	Mils
Centimeters	10	Millimeters
Centimeter-dynes	1.020×10^{-3}	Centimeter-grams
Centimeter-dynes	1.020×10^{-8}	Meter-kilograms
Centimeter-dynes	7.376×10^{-8}	Pound-feet
Centimeter-grams	980.7	Centimeter-dynes
Centimeter-grams	10^{-5}	Meter-kilograms
Centimeter-grams	7.233×10^{-5}	Pound-feet
Centimeters of mercury	0.01316	Atmospheres
Centimeters of mercury	0.4461	Feet of water
Centimeters of mercury	136.0	Kilograms/square meter
Centimeters of mercury	27.85	Pounds/square foot
Centimeters of mercury	0.1934	Pounds/square inch
Centimeters/second	1.969	Feet/minute
Centimeters/second	0.03281	Feet/second
Centimeters/second	0.036	Kilometers/hour
Centimeters/second	0.6	Meters/minute
Centimeters/second	0.02237	Miles/hour
Centimeters/second	3.728×10^{-4}	Miles/minute
Centimeters/second/second	0.03281	Feet/second/second
Centimeters/second/second	0.036	Kilometers/hour/second
Centimeters/second/second	0.02237	Miles/hour/second
Circular mils	5.067×10^{-6}	Square centimeters
Circular mils	7.854×10^{-7}	Square inches
Circular mils	0.7854	Square mils
Cord-feet	4 feet x 4 feet x 1 foot	Cubic feet

142 FUNDAMENTALS OF WIND ENERGY

Multiply	By	To Obtain
Cords	8 feet x 4 feet x 4 feet	Cubic feet
Cubic centimeters	3.531×10^{-5}	Cubic feet
Cubic centimeters	6.102×10^{-2}	Cubic inches
Cubic centimeters	10^{-6}	Cubic meters
Cubic centimeters	1.308×10^{-6}	Cubic yards
Cubic centimeters	2.642×10^{-4}	Gallons
Cubic centimeters	10^{-3}	Liters
Cubic centimeters	2.113×10^{-3}	Pints (liquid)
Cubic centimeters	1.057×10^{-3}	Quarts (liquid)
Cubic centimeters	0.033814	Ounces (U.S. fluid)
Cubic feet	2.832×10^{4}	Cubic centimeters
Cubic feet	1728	Cubic inches
Cubic feet	0.02832	Cubic meters
Cubic feet	0.03704	Cubic yards
Cubic feet	7.481	Gallons
Cubic feet	28.32	Liters
Cubic feet	59.84	Pints (liquid)
Cubic feet	29.92	Quarts (liquid)
Cubic feet of water (60°F)	62.37	Pounds
Cubic feet/minute	472.0	Cubic centimeters/second
Cubic feet/minute	0.1247	Gallons/second
Cubic feet/minute	0.4720	Liters/second
Cubic feet/minute	62.4	Pounds of water/minute
Cubic feet/second	1.9834	Acre-feet/day
Cubic feet/second	448.83	Gallons/minute
Cubic feet/second	0.64632	Million gallons/day
Cubic feet-atmospheres	2.7203	Btu (mean)
Cubic foot-atmospheres	680.74	Calories, gram (mean)
Cubic foot-atmospheres	2116.3	Foot-pounds
Cubic foot-atmospheres	292.6	Kilogram-meters
Cubic foot-atmospheres	7.968×10^{-4}	Kilowatt-hours
Cubic inches	16.39	Cubic centimeters
Cubic inches	5.787×10^{-4}	Cubic feet
Cubic inches	1.639×10^{-5}	Cubic meters
Cubic inches	2.143×10^{-5}	Cubic yards
Cubic inches	4.329×10^{-3}	Gallons
Cubic inches	1.639×10^{-2}	Liters
Cubic inches	0.03463	Pints (liquid)
Cubic inches	0.01732	Quarts (liquid)
Cubic inches (U.S.)	0.55411	Ounces (U.S. fluid)

CONVERSION FACTORS 143

Multiply	By	To Obtain
Cubic meters	10^6	Cubic centimeters
Cubic meters	35.31	Cubic feet
Cubic meters	61,023	Cubic inches
Cubic meters	1.308	Cubic yards
Cubic meters	264.2	Gallons
Cubic meters	10^3	Liters
Cubic meters	2113	Pints (liquid)
Cubic meters	1057	Quarts (liquid)
Cubic meters	8.1074×10^{-4}	Acre-feet
Cubic meters	8.387	Barrels (U.S., liquid)
Cubic yards (British)	0.9999916	Cubic yards (U.S.)
Cubic yards	7.646×10^5	Cubic centimeters
Cubic yards	27	Cubic feet
Cubic yards	46.656	Cubic inches
Cubic yards	0.7646	Cubic meters
Cubic yards	202.0	Gallons
Cubic yards	764.6	Liters
Cubic yards	1616	Pints (liquid)
Cubic yards	807.9	Quarts (liquid)
Cubic yards/minute	0.45	Cubic feet/second
Cubic yards/minute	3.367	Gallons/second
Cubic yards/minute	12.74	Liters/second
Days	1440	Minutes
Days	86,400	Seconds
Decigrams	0.1	Grams
Deciliters	0.1	Liters
Decimeters	0.1	Meters
Degrees (angle)	60	Minutes
Degrees (angle)	0.01745	Radians
Degrees (angle)	3600	Seconds
Degrees/second	0.01745	Radians/second
Degrees/second	0.1667	Revolutions/minute
Degrees/second	0.002778	Revolutions/second
Dekagrams	10	Grams
Dekaliters	10	Liters
Dekameters	10	Meters
Drams	1.772	Grams
Drams	0.0625	Ounces
Dynes	1.020×10^{-3}	Grams
Dynes	7.233×10^{-5}	Poundals
Dynes	2.248×10^{-6}	Pounds

144 FUNDAMENTALS OF WIND ENERGY

Multiply	By	To Obtain
Dynes per square centimeter	1	Bars
Ergs	9.486×10^{-11}	British thermal units
Ergs	1	Dyne-centimeters
Ergs	7.376×10^{-8}	Foot-pounds
Ergs	1.020×10^{-3}	Gram-centimeters
Ergs	10^{-7}	Joules
Ergs	2.390×10^{-11}	Kilogram-calories
Ergs	1.020×10^{-8}	Kilogram-meters
Ergs/second	5.692×10^{-9}	British thermal units/minute
Ergs/second	4.426×10^{-6}	Foot-pounds/minute
Ergs/second	7.376×10^{-8}	Foot-pounds/second
Ergs/second	1.341×10^{-10}	Horsepower
Ergs/second	1.434×10^{-9}	Kilogram-calories/minute
Ergs/second	10^{-10}	Kilowatts
Fathoms	6	Feet
Feet	30.48	Centimeters
Feet	12	Inches
Feet	0.3048	Meters
Feet	1/3	Yards
Feet (U.S.)	1.893939×10^{-4}	Miles (statute)
Feet of air (1 atmosphere 60°F)	5.30×10^{-4}	Pounds/square inch
Feet of water	0.02950	Atmospheres
Feet of water	0.8826	Inches of mercury
Feet of water	304.8	Kilograms/square meter
Feet of water	62.43	Pounds/square foot
Feet of water	0.4335	Pounds/square inch
Feet/minute	0.5080	Centimeters/second
Feet/minute	0.01667	Feet/second
Feet/minute	0.01829	Kilometers/hour
Feet/minute	0.3048	Meters/minute
Feet/minute	0.01136	Miles/hour
Feet/second	30.48	Centimeters/second
Feet/second	1.097	Kilometers/hour
Feet/second	0.5921	Knots/hour
Feet/second	18.29	Meters/minute
Feet/second	0.6818	Miles/hour
Feet/second	0.01136	Miles/minute

CONVERSION FACTORS 145

Multiply	By	To Obtain
Feet/100 feet	1	Percent Grade
Feet/second/second	30.48	Centimeters/second/second
Feet/second/second	1.097	Kilometers/hour/second
Feet/second/second	0.3048	Meters/second/second
Feet/second/second	0.6818	Miles/hour/second
Foot-poundals	3.9951×10^{-5}	Btu (mean)
Foot-poundals	0.0421420	Joules (abs)
Foot-pounds	0.013381	Liter-atmospheres
Foot-pounds	3.7662×10^{-4}	Watt-hours (abs)
Foot-pounds	1.286×10^{-3}	British thermal units
Foot-pounds	1.356×10^{7}	Ergs
Foot-pounds	5.050×10^{-7}	Horsepower-hours
Foot-pounds	1.356	Joules
Foot-pounds	3.241×10^{-4}	Kilogram-calories
Foot-pounds	0.1383	Kilogram-meters
Foot-pounds	3.766×10^{-7}	Kilowatt-hours
Foot-pounds/minute	1.286×10^{-3}	British thermal units/minute
Foot-pounds/minute	0.01667	Foot-pounds/second
Foot-pounds/minute	3.030×10^{-5}	Horsepower
Foot-pounds/minute	3.241×10^{-4}	Kilogram-calories/minute
Foot-pounds/minute	2.260×10^{-5}	Kilowatts
Foot-pounds/second	7.717×10^{-2}	British thermal units/minute
Foot-pounds/second	1.818×10^{-3}	Horsepower
Foot-pounds/second	1.945×10^{-2}	Kilogram-calories/minute
Foot-pounds/second	1.356×10^{-3}	Kilowatts
Foot-pounds/second	4.6275	Btu (mean)/hour
Foot-pounds/second	1.35582	Watts (abs)
Gallons (British)	4516.086	Cubic centimeters
Gallons (British)	1.20094	Gallons (KU
Gallons (British)	10	Pounds (avordupois) of of water at 62°F
Gallons (U.S.)	128	Ounces (U.S. fluid)
Gallons	3785	Cubic centimeters
Gallons	0.1337	Cubic feet
Gallons	231	Cubic inches
Gallons	3.785×10^{-3}	Cubic meters
Gallons	4.951×10^{-3}	Cubic yards
Gallons	3.785	Liters

Multiply	By	To Obtain
Gallons	8	Pints (liquid)
Gallons	4	Quarts (liquid)
Gallons/minute	2.228×10^{-3}	Cubic feet/second
Gallons/minute	0.06308	Liters/second
Grains (troy)	1	Grains (average)
Grains (troy)	0.06480	Grams
Grains (troy)	0.04167	Pennyweights (troy)
Grains (troy)	2.0833×10^{-3}	Ounces (troy)
Grains/U.S. gallons	17.118	Parts/million
Grains/U.S. gallons	142.86	Pounds/million gallons
Grains/Imperial gallons	14.286	Parts/million
Grams	980.7	Dynes
Grams	15.43	Grains (troy)
Grams	10^{-3}	Kilograms
Grams	10^{3}	Milligrams
Grams	0.03527	Ounces
Grams	0.03215	Ounces (troy)
Grams	0.07093	Poundals
Grams	2.205×10^{-3}	Pounds
Gram-calories	3.968×10^{-3}	British thermal units
Gram-centimeters	9.302×10^{-8}	British thermal units
Gram-centimeters	980.7	Ergs
Gram-centimeters	7.233×10^{-5}	Foot-pounds
Gram-centimeters	9.807×10^{-5}	Joules
Gram-centimeters	2.344×10^{-8}	Kilogram-calories
Gram-centimeters	10^{-5}	Kilogram-meters
Gram-centimeters	2.7241×10^{-8}	Watt-hours
Gram-centimeters/second	9.80665×10^{-5}	Watts (abs)
Grams-centimeters^2 (moment of inertia)	3.4172×10^{-4}	Pounds-inch^2
Grams-centimeters^2	2.37305×10^{-6}	Pounds-feet^2
Grams/cubic meters	0.43700	Grains/cubic foot
Grams/centimeter	5.600×10^{-3}	Pounds/inch
Grams/cubic centimeter	62.43	Pounds/cubic foot
Grams/cubic centimeter	0.03613	Pounds/cubic inch
Grams/cubic centimeter	3.405×10^{-7}	Pounds/mil foot
Grams/cubic centimeter	8.34	Pounds/gallon
Grams/liter	58.417	Grains/gallon (U.S.)
Grams/liter	9.99973×10^{-4}	Grams/cubic centimeter
Grams/liter	1000	Parts/million (ppm)

Multiply	By	To Obtain
Grams/liter	0.06243	Pounds/cubic foot
Grams/square centimeter	0.0142234	Pounds/square inch
Hectograms	100	Grams
Hectoliters	100	Liters
Hectometers	100	Meters
Hectowatts	100	Watts
Hemispheres (sol. angle)	0.5	Sphere
Hemispheres (sol. angle)	4	Spherical right angles
Hemispheres (sol. angle)	6.283	Steradians
Horsepower	42.44	British thermal units/minute
Horsepower	33,000	Foot-pounds/minute
Horsepower	550	Foot-pounds/second
Horsepower	1.014	Horsepower (metric)
Horsepower	10.70	Kilogram calories/minute
Horsepower	0.7457	Kilowatts
Horsepower	745.7	Watts
Horsepower (boiler)	33,520	British thermal units/hour
Horsepower (boiler)	9.804	Kilowatts
Horsepower, electrical	1.0004	Horsepower
Horsepower (metric)	0.98632	Horsepower
Horsepower-hours	2547	British thermal units
Horsepower-hours	1.98×10^6	Foot-pounds
Horsepower-hours	2.684×10^6	Joules
Horsepower-hours	641.7	Kilogram-calories
Horsepower-hours	2.737×10^5	Kilogram-meters
Horsepower-hours	0.7457	Kilowatt-hours
Hours	60	Minutes
Hours	3600	Seconds
Inches	2.540	Centimeters
Inches	10^3	Mils
Inches of mercury	0.03342	Atmospheres
Inches of mercury	1.133	Feet of water
Inches of mercury	0.0345	Kilograms/square centimeters
Inches of mercury	345.3	Kilograms/square meter
Inches of mercury	25.40	Millimeters of mercury

148 FUNDAMENTALS OF WIND ENERGY

Multiply	By	To Obtain
Inches of mercury	70.73	Pounds/square foot
Inches of mercury	0.4912	Pounds/square inch
Inches of water	0.002458	Atmospheres
Inches of water	0.07355	Inches of mercury
Inches of water	25.40	Kilograms/square meter
Inches of water	0.5781	Ounces/square inch
Inches of water	5.204	Pounds/square foot
Inches of water	0.03613	Pounds/square inch
Kilograms	980,665	Dynes
Kilograms	10^3	Grams
Kilograms	70.93	Poundals
Kilograms	2.2046	Pounds
Kilograms	1.102×10^{-3}	Tons (short)
Kilogram-calories	3.968	British thermal units
Kilogram-calories	3086	Foot-pounds
Kilogram-calories	1.558×10^{-3}	Horsepower-hours
Kilogram-calories	426.6	Kilogram-meters
Kilogram-calories	1.162×10^{-3}	Kilowatt-hours
Kilogram-calories/minute	51.43	Foot-pounds/second
Kilogram-calories/minute	0.09351	Horsepower
Kilogram-calories/minute	0.06972	Kilowatts
Kilogram-centimeters^2	2.373×10^{-3}	Pounds-feet^2
Kilogram-centimeters^2	0.3417	Pounds-inches^2
Kilogram-meters	9.302×10^{-3}	British thermal units
Kilogram-meters	9.807×10^7	Ergs
Kilogram-meters	7.233	Foot-pounds
Kilogram-meters	3.6529×10^{-6}	Horsepower-hours
Kilogram-meters	9.579×10^{-6}	Pounds water evaporated at 212°F
Kilogram-meters	9.807	Joules
Kilogram-meters	2.344×10^{-3}	Kilogram-calories
Kilogram-meters	2.724×10^{-6}	Kilowatt-hours
Kilograms/cubic meter	10^{-3}	Grams/cubic meter
Kilograms/cubic meter	0.06243	Pounds/cubic foot
Kilograms/cubic meter	3.613×10^{-5}	Pounds/cubic inch
Kilograms/cubic meter	3.405×10^{-10}	Pounds/mil foot
Kilograms/meter	0.6720	Pounds/foot

CONVERSION FACTORS 149

Multiply	By	To Obtain
Kilograms/square centimeter	28.96	Inches of mercury
Kilograms/square centimeter	735.56	Millimeters of mercury
Kilograms/square centimeter	14.22	Pounds/square inch
Kilograms/square meter	9.678×10^{-5}	Atmospheres
Kilograms/square meter	3.281×10^{-3}	Feet of water
Kilograms/square meter	2.896×10^{-3}	Inches of mercury
Kilograms/square meter	0.07356	Millimeters of mercury at 0°C
Kilograms/square meter	0.2048	Pounds/square foot
Kilograms/square meter	1.422×10^{-3}	Pounds/square inch
Kilograms/square millimeter	10^6	Kilograms/square meter
Kiloliters	10^3	Liters
Kilometers	10^5	Centimeters
Kilometers	3281	Feet
Kilometers	10^3	Meters
Kilometers	0.6214	Miles
Kilometers	1093.6	Yards
Kilometers/hour	27.78	Centimeters/second
Kilometers/hour	54.68	Feet/minute
Kilometers/hour	0.9113	Feet/second
Kilometers/hour	0.5396	Knots/hour
Kilometers/hour	16.67	Meters/minute
Kilometers/hour	0.6214	Miles/hour
Kilometers/hour/second	27.78	Centimeters/second/second
Kilometers/hour/second	0.9113	Feet/second/second
Kilometers/hour/second	0.2778	Meters/second/second
Kilometers/hour/second	0.6214	Miles/hour/second
Kilometers/minute	60	Kilometers/hour
Kilowatts	56.92	British thermal units/minute
Kilowatts	4.425×10^4	Foot-pounds/minute
Kilowatts	737.6	Foot-pounds/second
Kilowatts	1.341	Horsepower

Multiply	By	To Obtain
Kilowatts	14.34	Kilogram-calories/minute
Kilowatts	10^3	Watts
Kilowatt-hours	3415	British thermal units
Kilowatt-hours	2.655×10^6	Foot-pounds
Kilowatt-hours	1.341	Horsepower hours
Liters	10^3	Cubic centimeters
Liters	0.03531	Cubic feet
Liters	61.02	Cubic inches
Liters	10^{-3}	Cubic meters
Liters	1.308×10^{-3}	Cubic yards
Liters	0.2642	Gallons
Liters	2.113	Pints (liquid)
Liters	1.057	Quarts (liquid)
Liters/minute	5.885×10^{-4}	Cubic feet/second
Liters/minute	4.403×10^{-3}	Gallons/second
$\text{Log}_{10} N$	2.303	$\text{Log}_E N$ or $\text{Ln } N$
Log N or Ln N	0.4343	$\text{Log}_{10} N$
Meters	100	Centimeters
Meters	3.2808	Feet
Meters	39.37	Inches
Meters	10^{-3}	Kilometers
Meters	10^3	Millimeters
Meters	1.0936	Yards
Meters	10^{10}	Angstrom units
Meters	6.2137×10^4	Miles
Meter-kilograms	9.807×10^7	Centimeter-dynes
Meter-kilograms	10^5	Centimeter-grams
Meter-kilograms	7.233	Pound-feet
Meters/minute	1.667	Centimeters/second
Meters/minute	3.281	Feet/minute
Meters/minute	0.05468	Feet/second
Meters/minute	0.06	Kilometers/hour
Meters/minute	0.03728	Miles/hour
Meters/second	196.8	Feet/minute
Meters/second	3.281	Feet/second
Meters/second	3.6	Kilometers/hour
Meters/second	0.06	Kilometers/minute
Meters/second	2.237	Miles/hour
Meters/second	0.03728	Miles/minute
Meters/second/second	3.281	Feet/second/second

CONVERSION FACTORS 151

Multiply	By	To Obtain
Meters/second/second	3.6	Kilometers/hour/second
Meters/second/second	2.237	Miles/hour/second
Micrograms	10^{-6}	Grams
Microliters	10^{-6}	Liters
Microns	10^{-6}	Meters
Miles	1.609×10^5	Centimeters
Miles	5280	Feet
Miles	1.6093	Kilometers
Miles	1760	Yards
Miles (int. Nautical)	1.852	Kilometers
Miles/hour	44.70	Centimeters/second
Miles/hour	88	Feet/minute
Miles/hour	1.467	Feet/second
Miles/hour	1.6093	Kilometers/hour
Miles/hour	26.82	Meters/minute
Miles/hour/second	44.70	Centimeters/second/second
Miles/hour/second	1.467	Feet/second/second
Miles/hour/second	1.6093	Kilometers/hour/second
Miles/hour/second	0.4470	Meters/second/second
Miles/minute	2682	Centimeters/second
Miles/minute	88	Feet/second
Miles/minute	1.6093	Kilometers/minute
Miles/minute	60	Miles/hour
Milliers	10^3	Kilograms
Milligrams	10^{-3}	Grams
Milliliters	10^{-3}	Liters
Millimeters	0.1	Centimeters
Millimeters	0.03937	Inches
Millimeters	39.37	Mils
Millimeters of mercury	0.0394	Inches of mercury
Millimeters of mercury	1.3595^{-3}	Kilograms/square centimeter
Millimeters of mercury	0.01934	Pounds/square inch
Mils	0.002540	Centimeters
Mils	10^{-3}	Inches
Mils	25.40	Microns
Minutes (angle)	2.909×10^{-4}	Radians
Minutes (angle)	60	Seconds (angle)
Months	30.42	Days
Months	730	Hours

152 FUNDAMENTALS OF WIND ENERGY

Multiply	By	To Obtain
Months	43,800	Minutes
Months	2.628×10^6	Seconds
Myriagrams	10	Kilograms
Myriameters	10	Kilometers
Myriawatts	10	Kilowatts
Ounces	16	Drams
Ounces	437.5	Grains
Ounces	28.35	Grams
Ounces	0.0625	Pounds
Ounces (fluid)	1.805	Cubic inches
Ounces (fluid)	0.02957	Liters
Ounces (U.S. fluid)	29.5737	Cubic centimeters
Ounces (U.S. fluid)	1/128	Gallons (U.S.)
Ounces (troy)	480	Grains (troy)
Ounces (troy)	31.10	Grams
Ounces (troy)	20	Pennyweights (troy)
Ounces (troy)	0.08333	Pounds (troy)
Ounces/square inch	0.0625	Pounds/square inch
Parts/million	0.0584	Grains/U.S. gallon
Parts/million	0.7016	Grains/Imperial gallon
Parts/million	8.345	Pounds/million gallons
Pennyweights (troy)	24	Grains (troy)
Pennyweights (troy)	1.555	Grams
Pennyweights (troy)	0.05	Ounces (troy)
Pints (dry)	33.60	Cubic inches
Pints (liquid)	28.87	Cubic centimeters
Pints (U.S. liquid)	473.179	Cubic centimeters
Pints (U.S. liquid)	16	Ounces (U.S. fluid)
Poundals	13,826	Dynes
Poundals	14.10	Grams
Poundals	0.03108	Pounds
Pounds	444,823	Dynes
Pounds	7000	Grains
Pounds	453.6	Grams
Pounds	16	Ounces
Pounds	32.17	Poundals
Pound (troy)	0.8229	Pounds (av.)
Pounds (troy)	373.2418	Grams
Pounds of carbon to CO_2	14,544	British thermal units (mean)
Pound-feet (torque)	1.3558×10^7	Dyne-centimeters

CONVERSION FACTORS

Multiply	By	To Obtain
Pound-feet	1.356×10^7	Centimeters-dynes
Pound-feet	13,825	Centimeter-grams
Pound-feet	0.1383	Meter-kilograms
Pounds-feet^2	421.3	Kilogram-centimeters^2
Pounds-feet^2	144	Pounds-inches^2
Pounds-inches^2	2,926	Kilogram-centimeters^2
Pounds-inches^2	6.945×10^{-3}	Pounds-feet^2
Pounds of water	0.01602	Cubic feet
Pounds of water	27.68	Cubic inches
Pounds of water	0.1198	Gallons
Pounds of water evaporated at 212°F	970.3	British thermal units
Pounds of water/minute	2.699×10^{-4}	Cubic feet/second
Pounds/cubic foot	0.01602	Grams/cubic centimeter
Pounds/cubic foot	16.02	Kilograms/cubic meter
Pounds/cubic foot	5.787×10^{-4}	Pounds/cubic inch
Pounds/cubic foot	5.456×10^{-9}	Pounds/mil foot
Pounds/cubic inch	27.68	Grams/cubic centimeter
Pounds/cubic inch	2.768×10^4	Kilograms/cubic meter
Pounds/cubic inch	1728	Pounds/cubic foot
Pounds/cubic inch	9.425×10^{-6}	Pounds/mil foot
Pounds/foot	1.488	Kilograms/meter
Pounds/inch	178.6	Grams/centimeter
Pounds/square foot	0.01602	Feet of water
Pounds/square foot	4.882	Kilograms/square meter
Pounds/square foot	6.944×10^{-3}	Pounds/square inch
Pounds/square inch	0.06804	Atmospheres
Pounds/square inch	2.307	Feet of water
Pounds/square inch	2.036	Inches of mercury
Pounds/square inch	0.0703	Kilograms/square centimeter
Pounds/square inch	703.1	Kilograms/square meter
Pounds/square inch	144	Pounds/square foot
Pounds/square inch	70.307	Grams/square centimeter
Pounds/square inch	51.715	Millimeters of mercury at 0°C
Quadrants (angle)	90	Degrees
Quadrants (angle)	5400	Minutes
Quadrants (angle)	1.571	Radians
Quarts (dry)	67.20	Cubic inches

154 FUNDAMENTALS OF WIND ENERGY

Multiply	By	To Obtain
Quarts (liquid)	57.75	Cubic inches
Quarts (U.S. liquid)	0.033420	Cubic feet
Quarts (U.S. liquid)	32	Ounces (U.S. fluid)
Quarts (U.S. liquid)	0.832674	Quarts (British)
Radians	57.30	Degrees
Radians	3438	Minutes
Radians	0.637	Quadrants
Radians/second	57.30	Degrees/second
Radians/second	0.1592	Revolutions/second
Radians/second	9.549	Revolutions/minute
Radians/second/second	573.0	Revolutions/minute/minute
Radians/second/second	9.549	Revolutions/minute/second
Radians/second/second	0.1592	Revolutions/second/second
Revolutions	360	Degrees
Revolutions	4	Quadrants
Revolutions	6.283	Radians
Revolutions/minute	6	Degrees/second
Revolutions/minute	0.1047	Radians/second
Revolutions/minute	0.01667	Revolutions/second
Revolutions/minute/minute	1.745×10^{-3}	Radians/second/second
Revolutions/minute/minute	0.01667	Revolutions/minute/second
Revolutions/minute/minute	2.778×10^{-4}	Revolutions/second/second
Revolutions/second	360	Degrees/second
Revolutions/second	6.283	Radians/second
Revolutions/second	60	Revolutions/minute
Revolutions/second/second	6.283	Radians/second/second
Revolutions/second/second	3600	Revolutions/minute/minute
Revolutions/second/second	60	Revolutions/minute/minute
Seconds (angle)	4.848×10^{-6}	Radians
Spheres (solid angle)	12.57	Steradians
Spherical right angles	0.25	Hemispheres
Spherical right angles	0.125	Spheres

CONVERSION FACTORS

Multiply	By	To Obtain
Spherical right angles	1.571	Steradians
Square centimeters	1.973×10^5	Circular mils
Square centimeters	1.076×10^{-3}	Square feet
Square centimeters	0.1550	Square inches
Square centimeters	10^{-6}	Square meters
Square centimeters	100	Square millimeters
Square centimeters-centimeters squared	0.02420	Square inches-inches squared
Square feet	2.296×10^{-5}	Acres
Square feet	929.0	Square centimeters
Square feet	144	Square inches
Square feet	0.09290	Square meters
Square feet	3.587×10^{-8}	Square miles
Square feet	1/9	Square yards
Square feet-feet squared	2.074×10^4	Square inches-inches squared
Square inches	1.273×10^6	Circular mils
Square inches	6.452	Square centimeters
Square inches	6.944×10^{-3}	Square feet
Square inches	10^6	Square mils
Square inches	645.2	Square millimeters
Square inches (U.S.)	7.71605×10^{-4}	Square yards
Square inches-inches squared	41.62	Square centimeters-centimeters squared
Square kilometers	247.1	Acres
Square kilometers	10.76×10^6	Square feet
Square kilometers	10^6	Square meters
Square kilometers	0.3861	Square miles
Square kilometers	1.196×10^6	Square yards
Square meters	2.471×10^{-4}	Acres
Square meters	10.764	Square feet
Square meters	3.861×10^{-7}	Square miles
Square meters	1.196	Square yards
Square miles	640	Acres
Square miles	27.88×10^6	Square feet
Square miles	2.590	Square kilometers
Square miles	3.098×10^6	Square yards
Square millimeters	1.973×10^3	Circular mils
Square millimeters	0.01	Square centimeters
Square millimeters	1.550×10^{-3}	Square inches
Square mils	1.273	Circular mils

Multiply	By	To Obtain
Square mils	6.452×10^{-6}	Square centimeters
Square mils	10^{-6}	Square inches
Square yards	2.066×10^{-4}	Acres
Square yards	9	Square feet
Square yards	0.8361	Square meters
Square yards	3.228×10^{-7}	Square miles
Temperature (°C) + 273	1	Absolute temperature (°C)
Temperature (°C) + 17.8	1.8	Temperature (°F)
Temperature (°F) + 460	1	Absolute temperature (°F)
Temperature (°F) - 32	5/9	Temperature (°C)
Tons (long)	1016	Kilograms
Tons (long)	2240	Pounds
Tons (metric)	10^3	Kilograms
Tons (metric)	2205	Pounds
Tons (short)	907.2	Kilograms
Tons (short)	2000	Pounds
Tons (short)/square feet	9765	Kilograms/square meter
Tons (short)/square feet	13.89	Pounds/square inch
Tons (short)/square inch	1.406×10^6	Kilograms/square meter
Tons (short)/square inch	2000	Pounds/square inch
Watts	0.05692	British thermal units/minute
Watts	10^7	Ergs/second
Watts	44.26	Foot-pounds/minute
Watts	0.7376	Foot-pounds/second
Watts	1.341×10^{-3}	Horsepower
Watts	0.01434	Kilogram-calories/minute
Watts	10^{-3}	Kilowatts
Watt-hours	3.415	British thermal units
Watt-hours	2655	Foot-pounds
Watt-hours	1.341×10^{-3}	Horsepower-hours
Watt-hours	0.8605	Kilogram-calories
Watt-hours	367.1	Kilogram-meters
Watt-hours	10^{-3}	Kilowatt-hours
Weeks	168	Hours
Weeks	10,080	Minutes
Weeks	604,800	Seconds

CONVERSION FACTORS

Multiply	By	To Obtain
Yards	91.44	Centimeters
Yards	3	Feet
Yards	36	Inches
Yards	0.9144	Meters
Years (common)	365	Days
Years (common)	8760	Hours

REFERENCES

REFERENCES CITED

1. *The U.S. Food and Fiber Sector: Energy Use and Outlook, A Study of the Energy Needs of the Food Industry,* Economic Research Service, USDA for Subcommittee on Agricultural and Rural Electrification, Committee of Agriculture and Forestry, U.S. Senate, U.S. G.P.O., Washington, D.C. (September 1974).
2. Liljedahl, L. A. "Wind Energy Use in Rural and Remote Areas," *Wind Energy Conversion Systems, Second Workshop Proceedings,* MITRE Corp., MTR-6970, NSF-RA-N-75-050, F. R. Eldridge, Ed. (September 1975).
3. Institute of Gas Technology, "Detailed Energy Usage Profile in the Canton Test Homes," A.G.A. Project HC-4-14 (December 1973).
4. Ramakumar, R., H. J. Allison and W. L. Hughes. "Analysis of the Parallel-Bridge Rectifier System," IEEE Transactions on Industry Applications, Vol. IA-9, No. 4 (July/August 1973), pp. 425-436.
5. *Wind Energy Conversion Systems: Workshop Proceedings,* NSF/RA/W-73-006 (December 1973).
6. Templin, R. J. "An Estimate of Interaction of Windmills in Widespread Arrays," Technical Report LTR-LA-171, NRC (1974).
7. Olsson, L. E., O. Holme and R. Kreig. "Wind Characteristics and Wind Power Generation," *Wind Energy Conversion Systems, Second Workshop Proceedings,* MITRE Corp., MTR-6970, NSF-RA-N-75-050, F. R. Eldridge, Ed. (September 1975).
8. Justus, C. G. "Annual Power Output Potential for 100-kW and 1-MW Aerogenerators," *Wind Energy Conversion Systems, Second Workshop Proceedings,* MITRE Corp., MTR-6970, NSF-RA-N-75-050, F. R. Eldridge, Ed. (September 1975).

9. Stone, L. D., D. E. Jenne and J. M. Thorp. "Climatography of the Hanford Area," Battelle, BNWL-1604 (June 1972).
10. Crawford, K. C. and H. R. Hudson. "Behavior of Winds in the Lowest 1500 Ft. in Central Oklahoma," NOAA ERLTM-NSSC-48 (August 1970).
11. Glauert, H. *The Elements of Airfoil and Airscrew Theory* (New York: Cambridge University Press, 1959).
12. Kogan, A. and E. Nissim. "Shrouded Aerogenerator Design Study, Two Dimensional Shroud Performance," *Bulletin of the Research Council of Israel*, Vol. 11c (1962), pp. 67-88.
13. Robertson, J. and D. Ross. "Water Tunnel Diffuser Flow Studies," Pennsylvania State College, ORL Report No. 7958-143 (1949).
14. Kogan, A. and A. Seginer. "Final Report on Shroud Design," Department of Aeronautical Engineering, Technion, TAE, Report No. 32A (1963).
15. Sweeney, T. E. "The Princeton Windmill Program," Princeton University (April 1973).
16. Golding, E. W. "Studies of Wind Behaviour and Investigation of Suitable Sites for Wind-Driven Plants," *New Sources of Energy*, Proceedings of the United Nations Conference in Rome, Vol. 7 [GR/6(W)].
17. Golding, E. W. "The Generation of Electricity by Wind Power," Philosophical Library, New York (1956).
18. Cormier, Rene V. "An Annotated Listing—Tall Towers Instrumented for Wind Observations," *Technical Report 73-0179*, Air Force Cambridge Research Laboratory, L. G. Hanscom Field, Bedford, Massachusetts (March 1973).
19. Blackwell, B. F. and L. V. Feltz. "Wind Energy—A Revitalized Pursuit," Sandia Laboratories, SAND 75-0166 (March 1975).
20. Davidson, B. "Sites for Wind-Power Installations," Technical Note No. 63, WMO-No. 156. TP. 76, World Meteorological Organization (1964).
21. Frenkiel, J. "Wind Profiles Over Hills," *Quart, J. Royal Meteorol. Soc.* Vol. 88 (1963).
22. Plate, E. J. and C. M. Sheih. "Diffusion from a Continuous Point Source into the Boundary Layer Downstream from a Model Hill," Colorado State University (1965).
23. Scorer, R. S. "Mountain-Gap Winds: A Study of Surface Wind at Gibralter," *Quart. J. Royal Meteorol. Soc.* 78(335) (January 1952).
24. Plate, E. J. and Chi W. Lin. "The Velocity Field Downstream from a Two-Dimensional Model Hill," Fluid Dynamics and Diffusion Laboratory, Colorado State University (1965).
25. Briggs, J. "Airflow Around a Model of the Rock of Gibraltar," Meteorological Office Scientific Paper No. 18 (1963).

26. Garrison, J. A. and J. E. Cermak. "San Bruno Mountain Wind Investigation—A Wind Tunnel Model Study," Colorado State University (1968).
27. Corotis, R. B. "Statistical Analysis of Continuous Data Records," *J. Transp. Eng., Am. Soc. Chem. Eng.* 100 (TE1) (1974).
28. Lakla, R. F. "Frequency Functions of Wind Speed and Wind Direction for Air Pollution Studies," M.S. Thesis, Department of Civil Engineering, Northwestern University (1972).
29. Zlotnik, M. "Energy Storage for Wind Energy Conversion Systems," *Wind Energy Conversion Systems, Second Workshop Proceedings*, MITRE Corp., MTR-6970, NSF-RA-N-75-050, F. R. Eldridge, Ed. (September 1975).
30. Hottel, H. C. and J. B. Howard. *New Energy Technology—Some Facts and Assessments* (Cambridge: M.I.T. Press, 1971).
31. Eldridge, F. R. "Wind Machines," Report to NSF, Division of Advanced Energy and Resources, Grant No. AER-75-12937 (October 1975).
32. Merriam, M. F. "Wind Energy for Human Needs," *Technology Review* 79(3):29-39 (January 1977).

UNCITED REFERENCES AND SUGGESTED READING

Abbot, I. H. and A. E. Von Doenhoff. *Theory of Wing Sections* (New York: Dover Publications, Inc., 1948).

Adkins, B. and W. J. Gibbs. *Polyphase Commutator Machines* (New York: Cambridge University Press, 1951).

Allison, H. J., R. Ramakumar and W. L. Hughes. "A Field Modulated Frequency Down Conversion Power System," *IEEE Trans. on Industry Applications,* IA-9(2) (March/April 1973).

"Alternate Energy Sources in Hawaii," Hawaii Natural Energy Institute, University of Hawaii and Department of Planning and Economic Development, State of Hawaii (February 1975).

Archibald, P. B. "An Analysis of the Winds of Site 300 as a Source of Power," UCRL-51469, Lawrence Livermore Labs., University of California, Livermore, California (1973).

Ashley, H. and W. P. Rodden. "Wing-Body Aerodynamic Interaction," *Annual Review of Fluid Mechanics* 4 (1972).

Beebe, P. S. and J. E. Cermak. "Turbulent Flow Over a Wavy Boundary," Project THEMIS TR No. 16, Fluid Dynamics and Diffusion Lab, Colorado State University (1972).

Berke, B. L. and R. N. Meroney. *Energy from the Wind: Annotated Bibliography,* Colorado State University (August 1975).

Betz, A. "Energieumsetzungen in Venturidusen," *Naturwissenshaften* 10(S.3) (1929).

Bird, B. M. and J. Ridge. "Amplitude-Modulated Frequency Changer," *Proc. Inst. Elect. Eng.* 119(8) (August 1972).

Blackwell, B. F. and L. V. Feltz. "Wind Energy—A Revitalized Pursuit," Sandia Laboratories, SAND 75-0166 (March 1975).

Chen, L. T., E. O. Suciu and L. Morino. "Steady and Oscillatory Subsonic and Supersonic Aerodynamics Around Complex Configurations," *Am. Inst. Aero. Astro. J.* 13(3) (March 1973).

Chirgwin, K. M. and L. J. Stratton. "Variable-Speed Constant-Frequency Generator System for Aircraft," *AIEE Trans. on Industry Applications* (Part II), 78 (November 1959).

Chirgwin, K. M., L. J. Stratton and J. R. Toth. "Precise Frequency Power Generation from an Unregulated Shaft," *AIEE Trans.* 79 (January 1961).

Clark, W. "Winds of Change," *Smithsonian* (November 1973).

Clews, H. *Electric Power from the Wind*, Solar Wind Co. (1973).

Corotis, R. B., E. H. Vanmarcke and C. A. Cornell. "First Passage of Nonstationary Random Processes," *J. Eng. Mech. Dir., Am. Soc. Chem. Eng.* 98(EM2) (1972).

Donham, R. E. and W. P. Harvick. "Analysis of Stowed Rotor Aeroelastic Characteristics," *J. Amer. Helicopter Soc.* 12(2) (January 1967).

Eggers, A. J., Jr. *Solar Energy Program, Subpanel Report IX,* used in preparing the AEC Chairman's Energy Report to the President, NSF, WASH-1281-9 (November 1973).

Fales, E. N. "Windmills," in *Standard Handbook for Mechanical Engineers,* T. Baumeister and L. S. Marks, Eds. (New York: McGraw-Hill, 1970).

Glahn, H. R. "The Use of Decision Theory in Meteorology," *Monthly Weather Review* 92(9) (1964).

Golding, E. W. "Themods of Assessing the Potentialities of Wind Power on Different Scales of Utilization," *Solar and Aerolian Energy, Proceedings of the International Seminar at Sounion, Greece* (New York: Plenum Press, 1961).

Gross, A. T. H. "Wind Power Usage in Europe," Techtran Corp., N74-31534/2ST (August 1974).

Hall, M. G. "A Theory for the Core of a Leading Edge Vortex," *J. Fluid Mech.* 11 (1961).

Haugen, D. A., Ed. *Workshop on Micrometeorology* (Boston: American Meteorological Society, 1973).

Hoard, B. V. "Constant-Frequency Variable-Speed Frequency-Make-up

Generators," *AIEE Trans.—Applications and Industry Part II*, Vol. 78 (November 1959).

Hughes, W. *Design, Fabrication and Analysis of a Five Kilowatt, Continuous Duty Electrical Power System, Operating on Wind Energy*, NTIS PB239-272, Oklahoma State University (October 1974).

Hust, E. "Energy Use for Food," ORNAL-NSF-EP-57, Washington, D.C., NSF (1974).

Kogan, A. and A. Seginer. "Shrouded Aerogenerator, Design Study II, Axisymmetric Shroud Performance," Department of Aeronautical Engineering, Technion, TAE Report No. 32 (1963).

Kogan, A. and A. Seginer. "Final Report on Shroud Design," Department of Aeronautical Engineering, Technion, TAE Report No. 32A (1963).

Kostenko, M. P. "A.C. Commutator Generation with Frequency Regulation Independent of Rotational Speed," *Electrichestuo* No. 2 (1948).

Kostenko, M. P. and L. Piotvousky. *Electrical Machines*, Vol. II (Moscow: M.I.R. Publishers, 1969).

Lindsley, E. F. "Windpower," *Popular Sci.* (July 1974).

McCaull, J. "Windmills," *Environment* 15 (1973).

McCroskey, W. J. "Inviscid Flowfield of an Unsteady Airflow," *Am. Inst. Aero. Astro. J.* 11(2) (August 1973).

Mercier, J. A. "Power Generating Characteristics of Savonius Rotors," Davidson Lab. Letter Report 1181, Stevens Institute of Technology, New Jersey (November 1966).

Morino, L. and C. C. Kuo. "Subsonic Potential Aerodynamics for Complex Configurations: A General Theory," *Am. Inst. Aero. Astro. J.* 12(2) (February 1974).

Owen, T. B. "Variable-Speed Constant-Frequency Devices: A Survey of the Methods in Use and Proposed," *AIEE Trans* (Part II—Applications and Industry) 78 (November 1959).

Paice, D. A. "Speed Control of Large Induction Motors by Thyristor Converters," *IEEE Trans.* IGA 5:545 (October 1969).

Pigeaud, F. D. *Method of Calculation of Annual Overall Efficiency of Modern Wind-Power Plants*, Linguistic Systems, Inc., N74-15748/8 (November 1971).

Pigeaud, F. D. and R. Wailes. "Power in the Wind," *New Scientist*, London (May 1965).

Puthoff, R. L. and P. J. Sirocky. "Preliminary Design of a 100-kW Wind Turbine Generator," NASA TMX-71585 (August 1974).

Putnam, P. C. *Power from the Wind*. (New York: D. Van Nostrand, Pub., 1948).

Reed, J. W. "Wind Power Climatology," *Weatherwise* (December 1974).

Reed, J. W. "Wind Power Climatology of the United States," Sandia Labs., SAND 74-0348 (June 1975).

Runyan, H. L. "Unsteady Lifting Surface Theory Applied to a Propeller and Helicopter Rotor," Ph.D. Thesis, Loughborough University of Technology, England (1973).

Savino, J. M. Ed. *Wind Energy Conversion Systems: First Workshop Proceedings*, NASA, PB-231 341/9 (December 1973).

Savino, J. M. *A Brief Summary of the Attempts to Develop Large Wind-Electric Generating Systems in the U.S.*, NASA-Lewis Research Center (August 1974).

Schwartz, J. H. *Batteries for Storage of Wind-Generated Energy*, NASA-Lewis Research Center (June 1973).

Serrin, J. "Mathematical Principles of Classical Fluid Mechanics," *Encyclopedia of Physics*, Vol. VIII/1, S. Flugge, Ed. (Berlin: Springer-Verlag, 1959), pp. 125-263.

Smeaton, J. "On the Construction and Effects of Windmill Sails," *The Philosophical Transactions of the Royal Society* 51, London (1759).

South, P. and R. S. Rangi. "A Wind Tunnel Investigation of a 14-Ft Diameter Vertical-Axis Windmill," National Research Council Report LTR-LA-1051 (September 1972).

Swanson, R. K., C. C. Johnson and R. T. Smith. *Wind-Power Development in the United States*, Southwest Research Institute (February 1974).

Templin, R. J. "Aerodynamic Performance Theory for the National Research Council Vertical-Axis Wind Turbine," National Research Council Report LTR-LA-160 (June 1974).

Thomas, P. H. "Electric Power from the Wind," Federal Power Commission (March 1975).

Thomas, R., R. Puthoff, J. Savino and W. Johnson. "Plans and Status of the NASA-Lewis Research Center Wind Energy Project," NASA TMX-71701 (October 1975).

Thwaites, B. *Incompressible Aerodynamics* (England: Oxford University Press, 1960).

"Vertical-Axis Wind Turbine Technology Workshop," *Proceedings*, Lyle Weatherholt, Ed., Albuquerque, New Mexico, Document SAND 76-5586 (May 1976).

Wallace, J. A., Jr., "An Annotated Bibliography of Meteorological Tower and Mast Studies," WB/BS-5, U.S. Dept. of Commerce, Environmental Science Services Administration, Environmental Data Services, Washington, D.C. (January 1967).

Wilson, R. E. and P. B. S. Lissaman. *Applied Aerodynamics of Wind Power Machines*, Oregon State University (May 1974).

Yen, J. T. *Tornado-Type Wind Energy System*, Tenth Intersociety Energy Conversion Eng. Conference IEEE Cat. No. 75 CHO 983-7 TAB (August 1975).

INDEX

Adriaenszoon, Jan 15
aeration
   diffused 40
   system 37
aerodynamic analysis 115
aerodynamic considerations 57
aerogenerator 75,77
   efficiency 76
   power density 77
aesthetics 123
agricultural energy
   consumption 31
   requirements 30
   uses 29
air compressor 41
air pollution problems 119
airport siting 87
air streamlines 71
anemometer 53,63,88
   hot wire 53
   pressure plate 53
   sonic 53
aquatic foodchain 38,39
Archimedean screw 17,18
atmospheric boundary layer 56,89
auxiliary power generating
     system 81
aviation traffic 120
axis of rotation 58

back-up units 81
batteries 81,106
   automobile 104
   fused salt electrolytic 105
   lead-cobalt 104
   organic-electrolyte 105
   sodium-sulfur 105
   storage 104
bending loads 59
Betz
   coefficient 73
   limit 57
blade 59
   configuration 60
   counter-rotating 61
   high-speed double 74
   multi- 41,74
   variable pitch 75
   -tip-speed 73
blowers 41
   centrifugal 41
breeze 1
Burnham, John 27

Canada 10,12
Canadian government 9
capital cost 113
climate, cold 121

climatography 92
climatological distribution 85
compressed air 42,81,99
   storage system 42,106,107
compressed gas 41
   fuels 81
constant voltage pumping 106
contour map 7
conversion factors 139
Coriolis forces 119
cost 79,83
cross-sectional area 50
cross wind paddle
   design 68
   rotor 67
cross wind Savonius 67
   design 68
Cubitt, William 21

Darrieus rotor 63,65,74
   arrangements 64
   turbine 44
data sources 85
definitions 69
Denmark 12
design 82,84
   characteristics 69
   components 80
destratification 40
differential systems 75
diffuser 61
   -type shroud 62
direct mechanical pumping 42
dissolved oxygen 40
drag devices 59,63
drainage windmill 18

economic factors 80
Edge, James 22
Edison cells 104

efficiency 57,75,79
electric power 4
electrical energy
   consumption 8,31
   generation 43,80
   generation systems 75
   pumps 42
electrolysis 108
electrolyte base 105
electrochemical energy storage 104
electrochemical reactions 104
energy
   consumption for rural residential heating 36
   consumption in the U.S. 6
   conversion site 90,96
   gain 82
   payback time 82
   products 101
   source 4
   storage 32,99,106
   storage systems 81,103,113
environmental
   considerations 99,115
   impact 81
   issues 122
ERDA 33

farm production electrical energy 32
farmland energy consumption 30
Federal Aviation Administration (FAA) 86
Federal Communications Commission (FCC) 118
flywheel 111
formulas, basic 69
fossil fuels 99
France 9,15

INDEX 167

frequency distribution 53
frequency generator 75
fuel cells 44
fuels
    combustible 34
    compressed gas 81
    fossil 99
    nuclear 99
fused salt electrolytic batteries
    105

Gaussian distribution 91
generation of electrical power
    42,43,80
generators 49
    aero- 75-77
    frequency 75
    gas-turbine electrical 44
    induction 46
    synchronous 45,77
    vortex 102,103,118
    water-pumping powered electric
        4
    wind, horizontal-axis 77
geographical regions 99
glossary 129
Golding 82
ground roughness 92
Grumman Aerospace Corporation
    102
gusting 52,58

Halladay, Daniel 24
head-on machines 59
horizontal-axis
    machine 73
    rotor 58,59,60,75
    rotor designs 60,61
    turbine 57,101
    wind generator 77

hydride/fuel cell 106
hydroelectric power 107
hydrogen 34,81,108,109
    electric approach 34,35
    storage 99
hypolimnion 40

ice shedding 121
induction generator 46
inductive coils 81
Institute of Gas Technology 33
instruments 53
irrigation 41

kinetic energy 49

land-use requirements 123
land wind 2
large-scale wind power design
    80,124
latent-heat storage 109-111
    materials 110
lead-cobalt batteries 104
Lee, Edmund 19
lift devices 59
load factor 83
loading
    conditions 84
    effects 83

magnetic storage 81
magnetite 111
magneto 53
magnus effect 66,67
magnus rotors 66
marine data 87
mathematical modeling 91,97
measurement 54

mechanical
  energy 75
  storage 81,111
Meikle, Andrew 21
mesotrophic 37
meteorological
  data 51,54,85,86
  prediction 97
  studies 10,53
methane gas 108,109
modulated frequency systems 46
momentum theory 72
motion of air 1
mountain lake aeration 40
mountain sites 94
municipal uses 35
municipal waste treatment 37

National Climatic Center 54,85
National Oceanic and Atmospheric
  Administration 54
National Weather Service 86,88
natural gas 109
navigation 120
Netherlands 10,11,13
noise level 118,119
North-Holland corn-mill 20
North Holland drainage mill 20
North Sea 13
Norway 13
Norwegian sailing ships 19
nuclear fission energy 43
nuclear fuels 99
numerical techniques 97

Occupational Safety Problems 119
ocean wind 2
offshore
  area requirements 123
  WECS arrangement 121
  winds 108

optimum
  height 53
  operating 75
  output performance 75
output-to-weight ratio 59
oxygen, dissolved 40

performance
  characteristics 69
  curves 74
Persian windmills 15,16
phase change 109
photosynthesis 40
photothermal energy 36
physical disturbances 117
potential energy storage 105
power
  capacity of wind 5
  coefficient 72
  density 50,79
  density curve 77
  density duration curve 76,78
  extraction 57,70
  harnessed 70
  output 50,72,78,80
  regulation 66
  wind response 66
pressure energy 70,71
pumped-hydro application 42
pumped storage 81

rectifier system 44
reserve power 34
rotor blade 74
rotor diameter 79
rotor swept area 79
rotors
  American multiblade 73
  crosswind horizontal-axis
    58,67
  crosswind paddle rotor 67

INDEX 169

Darrieus 63-65,74
Darrieus turbine 44
   downwind 59
   horizontal-axis 58-61,75
   magnus 66
   Savonius 62,64,74
   unshrouded 69
   vertical-axis 58,62,63
rotary units 75
roughness elements 54,92
Rural Electrification Administration 4

safety considerations 115
sailwind windmill 61
Savonius
   rotor 62,64,74
   S-shaped design 64
seasonal changes 2,81
sensible-heat storage 81,109,110
   materials 110
shaft horsepower 33
shroud 61
site selection 85,95
Smeaton, John 19
Smith-Putnam wind turbine 43
smock mill 17
sodium-sulfur batteries 105
space heating 44
speed-torque plot 76
static frequency changers 75
Stevin, Simon 10
storage 36
   batteries 104
   reservoirs 106
   systems 81,114
stored hydrogen 44
stratified shear flow 92
Sweden 13
Swedish State Power Board 9
synchronous generator 45,77

Technion 12
terrain 94
thermal energy storage 81,109
   systems 111,114
thermal pollution 8
thermal stratification 37,38,40
thermal zone 37
thermoelectric type machines 58
tip-speed ratio 76,77
tjasker 17
topography 56,88,89
tower shadowing 83
transducer 53
tropical air 1
turbine 101,117
   blade failure 120
   Darrieus rotor 44
   horizontal-axis 57,101
   multiblade 41
   rated speed 75
   vertical-axis 102
   wind 57,70
turbulence 56,89
   structure 54
turnaround efficiency 113

United Kingdom 8
United States energy consumption 5

velocity 49,70
   duration curves 51
vertical-axis
   machines 57
   rotor 58,62
   rotor systems 62,63
   turbines 102
vortex generator 102,103,118

Walsh-Healey Act 118

water-pumping powered electric
    generators 4
WECS 99,102
Weibull scale factors 56
Wheeler, Lawrence 24,27
wind
    characteristics 49
    circulation 95
    contours 86
    direction 3,5,83
    duration profiles 53,89
    energy conversion system
        (WECS) 8
    energy density distribution
        54,55
    energy facilities 9
    energy program 33
    energy site 94
    formation 1
    furnaces 36
    gust 53,54,83
    powered aeration system 41
    powered electrical generating
        4,45
    powered hydrogen 34
    profiles 56,94
    shear 91
    spectra 54
    speed 49,75,89
    speed data 56
    speed values 86
    stream influence 116
    tunnel tests 10
    turbine 37,118
    turbine efficiency 70
    variability 57
    velocity 52,55,75,79,88
    velocity measurement 53,54
wind machines 15,49,58
    performance 75
windmill 4,15,22,69
    American multiblade rotor
        21,23,74
    blades *see* blades
    drainage 18
    Dutch four-arm design 73,74
    industry 25,26
    Persian 109
    sailwind 61
self-regulating 25
World Meteorological Organization
    96